Handbook of Paleozoology

JOHNS HOPKINS STUDIES IN EARTH AND PLANETARY SCIENCES

Owen M. Phillips and Steven M. Stanley, Consulting Editors

Handbook of Paleozoology

Emil Kuhn-Schnyder PROF. DR. SC. NAT.

Hans Rieber PROF. DR. RER. NAT.

Paleontological Institute and Museum
of the University of Zurich

translated by
Emil Kucera PH.D. (Vertebrate Zoology)

THE JOHNS HOPKINS UNIVERSITY PRESS BALTIMORE AND LONDON

Originally published as *Paläozoologie: Morphologie und Systematik der ausgestorbenen Tiere,* © 1984 Georg Thieme Verlag, Stuttgart

The Johns Hopkins University Press
701 West 40th Street
Baltimore, Maryland 21211
The Johns Hopkins Press Ltd., London

LIBRARY OF CONGRESS CATALOGING-IN-PUBLICATION DATA
Kuhn-Schnyder, Emil.
  Handbook of paleozoology.
  (Johns Hopkins studies in earth and planetary sciences)
  Translation of: Paläozoologie.
  Bibliography: p.
  Includes indexes.
  1. Paleontology. I. Rieber, Hans. II. Title. III. Series.
QE761.K8413 1986   560   86-45454
ISBN 0-8018-2837-6 (alk. paper)

Contents

Foreword

This book aims to provide students with a well-illustrated, condensed introduction to the systematics of fossil animals. Considering the abundance of currently known animal fossils, it was necessary to make a strong, admittedly subjective, selection.

As paleontology is primarily a biological science, the higher systematic categories could not be described without reference to their living representatives. Modern paleontology must pay particular attention to the skeletons of living organisms, because it is mainly the hard, skeletal parts that remain preserved as fossils.

Current animal taxonomy is in an unsettled state. Although newer theories were not disregarded, this book generally observes a conservative classification. The condensed form precludes in-depth discussions of the precise systematic status of various groups as well as treatment of very small groups and of Precambrian fossils.

The constraints of space require abandoning much of the functional, ecological, and stratigraphic information, and detailed referencing of authors and sources. Those who wish to study the taxonomy in more depth will find a list of further reading at the end of the book. E. KUHN-SCHNYDER is responsible for the chapters on Chordata, and H. RIEBER for descriptions of the remaining phyla. All text illustrations were drawn by MR. O. GARRAUX; their origin is indicated in the captions. We wish to express our sincere thanks to MR. GARRAUX for his conscientious work. We are also grateful to the publishers for their helpful understanding during the execution of this work.

Zurich, Spring 1984   EMIL KUHN-SCHNYDER, HANS RIEBER

Handbook of Paleozoology

Introduction

The science of paleontology is concerned with the study of life through the geological history of the Earth (literally, a science of ancient organisms). Its goal is to reconstruct the history of life on Earth, and its means to achieve it is the evidence of fossils. As the study objects of paleontology are the remains of earlier living organisms, paleontology is accounted as one of the branches of biology, and the biological research methods apply to it as well. However, no experimental methods are open to paleontology.

Although the question of the origin of fossils has intrigued mankind since antiquity, paleontology as a science is comparatively new. It was not until the end of the seventeenth century that the organic origin of the fossils became generally recognized. The increasing knowledge of the fossil forms contributed significantly to the general knowledge of the diversity of life. Endeavors to gain an intellectual understanding of the seemingly overwhelming diversity of organisms became the central theme of biology during the eighteenth century. It was necessary to describe and classify the multitude of the species. CARL VON LINNÉ (Linnaeus, 1707–1778) created a classification system and introduced the binomial nomenclature. He was the first to note the increasing order of complexity of living things and to outline a natural system—the order in which God created them. Following Linnaeus, the brilliant GEORGES CUVIER (1769–1832) demonstrated that a natural system can be based on the findings of comparative anatomy. A great advancement was the discovery of CHARLES DARWIN (1809–1882) that the progressive complexity of organisms also increasingly reflects their genetic relationships (1859). The extant plants and animals have gradually evolved from geologically older species (transformism, evolution theory). The first grand outline of the phylogeny of animals is the work of ERNST HAECKEL (1834–1919). Although recently the theory of evolution has been much criticized, no existing paleontological find contradicts it.

In general, no fossils arise under natural circumstances, because dead organisms are either chemically decomposed or consumed by scavengers. Even their hard parts are usually converted to inorganic matter through mechanical and chemical processes. The minerals then become available for incorporation into new organisms. Fossils originate only when the natural decomposition is somehow prevented or interrupted. As it is mostly the hard parts (shells, skeletons, teeth) that are not completely decomposed, they make up the majority of fossils. The animal hard parts contain but few minerals, most commonly calcium carbonate (as calcite and aragonite), hydroxylapatite, and opaline silica.

The closer the relationship between the soft parts and the skeleton, the easier it is to deduce the structure and function of a fossil animal. Thus, of the fossil organisms without a skeleton, practically nothing is known, nor is it likely that much more will ever come to light. Yet they constitute more then two-fifths of all animal phyla.

Progress in paleontology is predicated on stratigraphically well-defined study material and on the researcher's commitment to constant refinement and improvement of both preparation techniques and research methods. Many of the present paleontological sites have been discovered accidentally. With the growing body of knowledge of the conditions in which fossils originate, it ought to be possible to locate promising fossil assemblages systematically.

The genealogy of animal remains is mostly imperfect and incomplete. Prerequisite to its determination is a thorough knowledge of the Recent animals. As opposed to zoology, paleontology cannot define species on the basis of reproductive and genetic relationships (biological species, biospecies) but only as morphological species, or morphospecies. Morphospecies encompass all those individuals that share important morphological characteristics. Paleontologists base their study of animal relationships on the comparison of homologous characteristics. Homologous characteristics in two or more organisms are derived from a common ancestor. However, similarity is not necessarily a proof of genetic relationship; it could also result from parallel or convergent evolution.

Paleontology contributes only chronologically ordered findings, as it cannot directly observe the speciation processes. Still, on the basis of the homology research, certain finds can be logically connected and ordered in succession. The succession lines give a broad outline of the probable course of the phylogeny. All the same, the resulting phylogenetic trees remain hypothetical, because the fossils are but indicators of the phylogeny. In many cases ontogenetic studies may be substituted for the missing historical evidence; it is, however, necessary to keep in mind that an embryo develops under conditions quite different from those encountered by its phylogenetic ancestor.

The assembled morphological data lead to physiological considerations. In many cases the embedding of fossils in sediments allows us to speculate about the environmental conditions during the time they were deposited. By applying the ecological requirements of Recent organisms to their fossil ancestors, the way of life of the latter can be surmised; and the composition of the fossil communities provides clues regarding their niche or at least their habitat.

Paleontology contributes to science in general in many ways:

1. Evidence of an immense multitude of fossil organisms.
2. Temporal distribution of living things, in connection with the evolution from simple to highly complex forms; a relative chronology of the last 530 million years of the Earth's history (stratigraphy).
3. Indication of evolutionary trends, based on the gradual development of characteristics over time.
4. Proof of the "missing links" and of the occurrence of "mosaic-evolution."
5. Proof of certain phases of accelerated evolution, based on the absolute chronology obtained by measuring the decay of radioactive elements.
6. Solution of zoogeographic problems, greatly aided by the Continental Drift Theory.
7. Information about the environment and the way of life of fossil organisms.

In addition to morphology, the taxonomy also utilizes other zoological disciplines (physiology, ecology, zoogeography, ethology, biochemistry, caryology, parasitology). Every classification system is thus a comprehensive reflection of contemporary knowledge.

Remarks

Measurements given in figure legends always relate to the individual subjects. Because the magnification factors were rounded off, the sizes may not be exact. In contrast, the stratigraphical data are always related to the taxa, mostly genera, as per the legends. The conception of individual genera is usually rather broad.

Major Division of Organisms

The traditional division of organisms into two superkingdoms, Procaryota and Eucaryota, is based on the structure and size of their
cells. Procaryota, with small cells, do not have a membrane-bound
nucleus, while the generally larger cells of the Eucaryota have a
nucleus separated from cytoplasm by a unit membrane. By the manner of their nutrition, Eucaryota can be further distinguished as plants
or animals (Plant and Animal kingdoms). Within the animal kingdom
there are two easily distinguishable subkingdoms: unicellular and
multicellular organisms (Protozoa and Metazoa). The plant kingdom
is divided into several more subkingdoms, according to the structure
of their representatives.

Against this traditional division, R. H. WHITTACKER (1969) recognizes
five kingdoms of organisms (Monera, Protista, Plantae, Fungi, and Animalia). Whittacker's division is based on three structural planes (pro-
caryotic, eucaryotic monocellular, and eucaryotic multicellular) and on
the feeding strategies. Although this classification has certain advantages
over the traditional one, the ranking of individual groups is still far from
simple.

Superkingdom **Procaryota**

Procaryota lack any hard, mineralized parts and are therefore rarely
preserved as fossils. However, by the stromatolites, they are already
known from the early Precambrian. Stromatolites, which also occur
today, are lumpy or dome-shaped inorganic bodies with a laminar
structure. They are formed by blue algae or bacterial associations as
they gradually bind fine sedimentary particles, predominantly calcium
carbonate, from their water environment.

Superkingdom **Eucaryota**

Kingdom **Animalia (animals)**

Uni- or multicellular heterotrophic organisms.

1. Subkingdom **Protozoa (protozoans, unicellular animals)**

Precambrian–Recent

The protozoan body functions morphologically as a cell and physiologically as an individual. The cell may contain one or more nuclei. Because it assigns specialized functions (such as locomotion, feeding, excretion, breathing, perception and transfer of stimuli, or protection) to specific parts of the cell, the organelles, a protozoan cell is an organism in its own right. Reproduction can be sexual, accompanied by fusion of the nuclei, or asexual, by division of the cell into two or more daughter organisms. The formation of protozoan colonies suggests a transition to the multicellular organisms.

Protozoans are mostly microscopic (2 micrometers to near 1 mm), yet there are also forms that can grow to 15 cm in size.

All protozoans are denizens of water or watery fluids. Thanks to their tiny size, many can exist with very little water. Most are heterotrophic, but among the Flagellata are autotrophic chromatophores—containing forms with plantlike metabolism. Many flagellates are mixotrophic. There are approximately 25,000 recent and 20,000 fossil protozoan species.

Phylum Protozoa

Of the four protozoan classes, only the Rhizopoda, with their many known fossil representatives, are well documented. It is true that the Flagellata class, too, has many fossils to offer, e.g., the Coccolithophorida (Jurassic–Recent), Silicoflagellata (Upper Cretaceous–Recent), and Dinoflagellata (Silurian, Permian–Recent), but these are autotrophic flagellates (Phytoflagellata). The tiny shells and skeletal remains of these flagellates are of greatest importance for stratigraphy and paleoclimatology. In addition to these nannofossils, which without doubt belong to the Flaggellata, the Acritarcha group (Precambrian–Recent) includes forms whose classification is not yet completely resolved.

Class Rhizopoda

? Precambrian, Cambrian–Recent

Protozoa whose protoplasm extends varying shapes of pseudopodia, for both locomotion and capture of prey. Many representatives of this

6

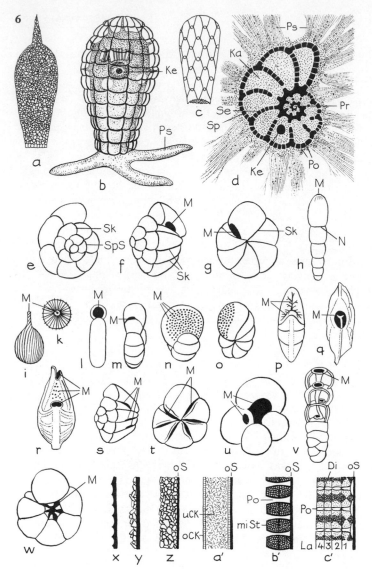

FIGURE 1. Class Rhizopoda. **a–c** Order Arcellinida. **d–c′** Order Foraminifera. **a** *Difflugia*, 120x, Recent, test agglutinated. **b** *Quadrulella*, 270x, Recent, test with chitinous platelets. **c** *Tracheleuglypha*, 400x, Recent, test with platelets of silica. **d** Schematic of a perforate foraminifer. **e–g** Trochoid foraminifer test from three aspects. **h** Uniserial test. **i′–w** Different types of appertures of the foraminifers. **x–c′** Shell structure. **x** Flexible tectine wall. **y** Tectine wall with loosely agglutinated particles. **z** Agglutinated wall. **a′** Thick, imperforate bilaminar calcite wall (porcelaneous type). **b′** Perforate microgranular calcareous wall. **c′** Perforate, multilaminar, microgranular calcareous wall.

group form supportive or protective organelles, of calcium carbonate, silica, or organic substances, with or without mineral inclusions.

Order Arcellinida (Thecamoebida)

Upper Carboniferous, Eocene–Recent (figure 1, a–c)

The representatives of this order are amoeboid, mostly fresh water– inhabiting protozoans with single-chambered shells. The shell (theca) measures from 0.5–0.2 mm and may consist of organic substances without or with agglutinated foreign material, of siliceous platelets, or, rarely, of calcium particles.

Certain Arcellinida are known from as early as the Eocene. The numerous Pleistocene forms are especially useful, as they allow ecological conclusion.

Order Foraminiferida

Cambrian–Recent (figures 1–61)

Foraminifers are mainly marine protozoans, with single- or unilocular (one chamber) or multilocular (several chambers) tests. Most of

Abbreviations used in figures 1–6

A	Antetheca	oS	organic inner layer
Aeka	equatorial chamber	Pf	pillars
AeS	equatorial section	Po	pores
ASe	section of parts of septa	Pr	proloculus
AxS	axial section	Ps	pseudopodia
Ch	choma	Se	septum
Di	diaphragm	SeF	septal filaments
Em	embryonal apparatus	SeP	septal pores
Ka	chamber	Sk	chamber suture
KaHo	chamber cavity	Sp	spirotheca
Ke	nucleus	SpS	spiral suture
Kk	chamberlet	Ss	septulum (secondary septum)
La	wall lamina	St	frontal wall (distal wall of
LKa	lateral chamber		the first chamber) with aper-
M	aperture		tures
miSt	microgranular wall structure	TaS	tangential section
N	suture	uCK	irregularly arranged calcite
Na	reticulopodia		crystals
NaPf	umbilical plug or column	Wa	warts
oCK	axially arranged calcite crys-	WAeka	wall of the equatorial cham-
	tals		ber

(after *Brasier, Cushman, Dunbar* and *Henbest, Kühn, Loeblich* and *Tappan, Pokorny, Reichel, Schaub*).

them range from 0.1–0.5 mm in diameter, but there are also foraminiferid tests from 5–15 cm in length or diameter. This makes them
the largest monocellular animals.

THE TEST (figure 1, d–c′)

Composition and morphology of the test are important taxonomic
characteristics.

The test may be composed of:

1. Tectine, a protein-polysaccharide mixture. Foraminifera with such mem-
   branaceous or "chitinose" tests are seldom preserved as fossils.
2. Agglutinated foreign particles (grains of sand, fine shell fragments,
   coccoliths, ooids, etc.) cemented to the test wall with tectine, silica, calcium, or ferruginous minerals. Depending on the nature of the cement,
   the organisms can have a high (silica) or low (tectine) potential for fossil
   preservation.
3. Tiny calcite particles, bonded together by a calcareous substance. The
   foraminifers with such microgranular wall structure are well capable of
   preservation.
4. The finest calcium carbonate crystallites of high- or low-magnesium calcite or, in some forms, of aragonite.
   a) The hyaline types have a thin, translucent wall, perforated by numer-
      ous pores. The calcium-carbonate crystallites may be either oriented
      with their C-axis perpendicular to the surface of the chamber (radial
      hyaline) or arranged irregularly (granular hyaline). In multilocular for-
      aminifers the new growth may, in addition, overlay all previous cham-
      bers, so that the wall of the older chambers appears multilaminar.
      When such a process does not take place, the walls may still consist of
      different layers, but without laminar structure.
   b) The porcelaneous types have an imperforate, relatively thick test wall.
      It is not translucent, and its high-magnesium calcite crystallites are
      either irregularly arranged or stacked bricklike.

Test Morphology

Tests of the unilocular foraminifers are often spherical or tubular.
Multilocular tests begin with an original chamber, called proloculus,
which is adjoined by additional chambers, separated by septa. The
joint of a septum and the test wall is frequently on the outside and
somewhat depressed; it is called a suture. The tests frequently assume
a plani- or trochospiral (domed spiral) shape. In addition to these,
tests that are stretched out in a line (uniserial), or involute tests,

occur; in these the newer chambers completely envelope the older ones. Irregular tests occur predominantly in sessile species.

The distal wall of the last-formed chamber always contains one or more apertures through which the cytoplasm can protrude. Their shape and position are taxonomically important because, together with the composition and morphology of the test wall, they are the most reliable characteristics allowing us to distinguish between the many similar, yet unrelated forms.

REPRODUCTION

To date the reproductive characteristics are known for only a few species. Usually a sexually produced generation (agamont, or schizont) and an asexually produced alternate generation (gamont) occur. Tests of both generations may either correspond or, to a greater or lesser degree, differ from each other. The diploid agamont (also called B-form) is designated microspheric because it generally has a small proloculus. Its adult test is generally much larger than the test of the haploid gamont. The haploid gamont, however, has a large proloculus (and is therefore designated macro- or megalospheric, or A-form). The macrospheric, asexually produced generation consistently outnumbers the microspheric one, from 2:1 to 30:1.

CLASSIFICATION (figures 2–6, table 1)

The five suborders were classified on the basis of the composition and microscopic structure of the test wall. Key taxonomic characteristics for the lower categories are primarily the number of chambers (uni- or multilocular tests), aperture characteristics, shape and order of chambers, manner of the perforation, and whether they are mobile or sessile.

Macroforaminifers (figures 2, e–g, 3, a, b, 4, 5, n–r, 6, g)

Especially in Upper Carboniferous and Permian, as well as from Upper Cretaceous into Mid-Tertiary, there were mass occurrences of giant foraminifers. Tests of the many different species reached from several millimeters up to about 12 cm (generally 0.5–2 cm) in length or diameter.

In warm, shallow seas these benthic-living giant foraminifers contributed to the formation of sedimentary rocks on a local to regional scale (fusulinid limestones of the early Paleozoic, orbitolinid limestones of the Upper Cretaceous, nummulitid and alveolinid lime-

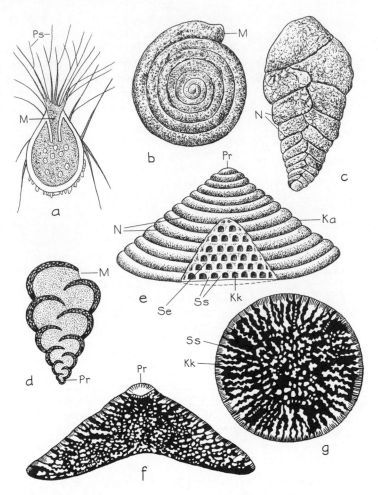

FIGURE 2. Order Foraminiferida.
a Allogromiina. **b–g** Textulariina.
a *Allogromia*, 30x, Recent.
b *Ammodiscus,* 90x, Upper Carboniferous. **c** *Textularia,* 60x, Pliocene. **d** *Textularia*—longitudinal section. **e** *Orbitolina* (schematic). **f, g** *Orbitolina* (longitudinal and cross section, 16x, Lower Cretaceous. For abbreviations see figure **1.**

stones of the late Tertiary). Descendants of the tertiary Alveolinidae live today in the coral reefs of tropical seas. Many of these disk-, lens-, or spindle-shaped macroforaminifers turned out to be excellent index fossils. Important in this respect are the genera *Fusulina,*

TABLE 1.   Overview of Suborders of the Foraminiferida

Suborder Geological Range (temporal distribution)	Shell Composition and Structure	Abundance and Variety of Forms	Important Fossil Forms	Figures
Allogromiina Upper Cambrian– Recent	Tectine with or without foreign particles; unilocular	fossils rare; low	—	2, a
Textulariina Lower Cambrian– Recent	Agglutinated foreign particles; uni- or multilocular	Fossils abundant; large	In addition to many smaller forms, there are macro-foraminifers (*Orbitolina*).	2, b–g
Fusulinina Ordovician– Triassic	Microgranular calcite; multilocular	Fossils abundant; large	*Fusulina, Triticites, Schwagerina*	3, a, b
Miliolina Carboniferous –Recent	Calcite, imperforate, thick, porcelaneous wall; multilocular	Fossils abundant; large	In addition to many smaller forms, there are macro-foraminifers (Alveolinidae).	3, c–f, 4
Rotaliina Permian– Recent	Mostly calcite, perforate, thin, hyaline; multilocular, frequently with a multilaminar wall	Fossils very abundant, very large	In addition to many smaller forms, there are also important macro-foraminifers (*Nummulites, Discocyclina*).	5, 6

Schwagerina, Alveolina, and *Nummulites* with planispiral tests, as well as *Orbitoides, Discocyclina,* and *Lepidocyclina* with circularly arranged chambers in the equatorial plane and additional lateral chambers. Many forms have a very complex interseptal canal system.

ECOLOGY

The overwhelming majority of foraminifers inhabit the sea. Most of them are vagile, benthic organisms, some are sessile, and a few are

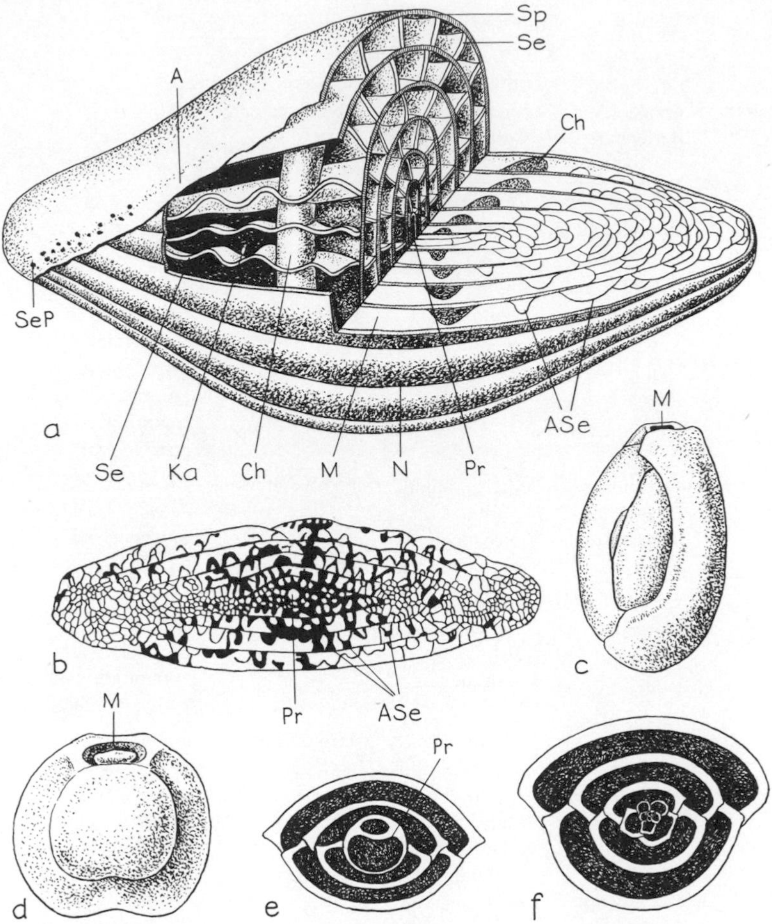

FIGURE 3. Order Foraminiferida. **a, b** Fusu-
linina. **c–f** Miliolina. **a** *Tricites*–test
(schematic), 40x, Upper Carboniferous.
b *Fusulina* (axial longitudinal section),
11x, Upper Carboniferous. **c** *Quin-*
queloculina, 30x, Eocene. **d–f** *Pyrgo*
(side view, macro- and microspheric
generation), 20x, Jurassic– Recent. For
abbreviations see figure **1**.

planktonic. The distribution, in particular of the benthic forms, is
very much affected by water temperature. Species that prefer colder
temperatures have less variable geographical distribution than do
warmth-loving ones. Thus, the benthic foraminifers can be used as

indicators of the different saltwater habitats as well as the depth or temperature zones. The very few (25 Recent) planktonic species occur in immense numbers. Sediments over vast areas of the ocean floor above the CCD (calcite compensation depth, below which calcite is dissolved) borderline consist of foraminiferid tests (globigerine ooze).

Planktonic and benthic foraminifers are guide (index) fossils with broad practical applications.

Fossil Foraminifera are also an excellent subject for phylogenetic study. Thanks to their great numbers, it is possible to work with large populations from different strata.

Order Radiolaria

? Precambrian, Cambrian–Recent (figure 7, a, b, table 2)

Radiolarians are protozoans distinguished by the presence of a central capsule. The central capsule is a porous, organic membrane dividing the intracapsular endoplasm from the outside ectoplasm. Most radiolarians measure from 0.1–0.5 mm. Their skeletons consist of amorphous silica or, less often, of strontium sulphate, and exhibit a great variety of often very elegant forms.

TABLE 2.   Overview of Suborders of the Radiolaria

Suborder Geological Range	Central Capsule	Skeleton	Composition	Figure
Acantharia Eocene– Recent	Perforated all over	Spiny sphere	Strontium sulphate	—
Spumellina Cambrian– Recent	Perforated all over	Lattice spheres, lattice disks	Opaline silica	7, a
Nasselina Cambrian– Recent	Only one porous field	Uniaxial form: cap- and cup-shaped	Opaline silica	7, b
Phaeodaria Cretaceous– Recent	With one main and two secondary openings	Often lattice spheres	Organic substance with approx. 5% silica	—

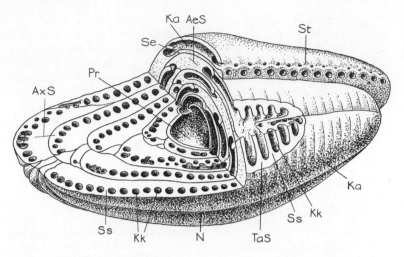

FIGURE 4. Order Foraminiferida: Miliolina. Schematic representation of an *Alveo-lina* test, 40x. Abbreviations as in figure **1**.

All radiolarians are planktonic, most of them occurring on the open seas, and there is a definite depth zonation with certain associations. Most individuals live in the upper 100 m of the ocean. Where the ocean water is high in nutrients, radiolarians may occur en masse.

Large areas of the ocean floor below the CCD borderline are covered with radiolarian ooze that consists primarily of Radiolaria tests. In the radiolarite, which is fossil radiolarian ooze, it is often difficult to distinguish the individual radiolarians.

Radiolaria are of increasing importance in paleoclimatic and stratigraphic research.

The four suborders of Radiolaria are based mainly on the skeletal pattern (which reflects the structure of the central capsule) and on its composition.

Class Ciliata

Ciliata characteristically have two kinds of nuclei. Their cell surface is either completely or partially covered by cilia. The coordinated movement of the cilia serves either for locomotion or to capture and transport food particles.

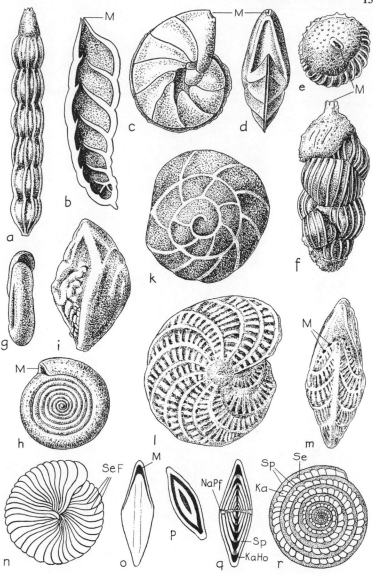

FIGURE 5. Order Foraminiferida: Rotaliina.
a *Nodosaria,* 26x, Cretaceous. **b** *Marginulina,* longitudinal section, 30x, Recent. **c, d** *Lenticulina,* 20x, Cretaceous. **e, f** *Uvigerina,* 80x, Pliocene. **g, h** *Spirillina,* 100x, Recent. **i, k** *Rotalia,* 20x, Eocene.

l, m *Elphidium,* 50x, Recent. **n–r** Schematic showing the structure of a *Nummulites* test, 3x, Eocene. **n** Side view of an eroded individual. **o** Frontal view. **p** Tangential section. **q** Median axial section. **r** Equatorial section. Abbreviations as in figure **1**.

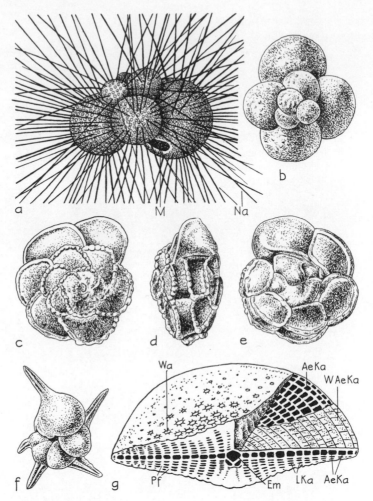

FIGURE 6. Order Foraminiferida: Rotaliina.
a *Globigerina*, 50x, Recent. **b** *Globigerina*, 70x, Recent. **c–e** *Glo-
botruncana*, 60x, Upper Cretaceous.

f *Hantkenina*, 25x, Eocene.
g Schematic of an orbitoid foraminiferid test, 15x, Eocene. Abbreviations as in figure **1**.

Only a few Recent Ciliata form permanent shells. The sea or
freshwater planktonic representatives of Tintinnidae (figure 7, c–e)
produce a bell-shaped lorica of organic substance in which foreign
particles may be embedded. The loricae measure between 70 and 400

microns. The few preserved fossils with organic-substance loricae may also belong to the Tintinnidae.

Protista incertae sedis: Calpionellidea

Upper Jurassic–Lower Cretaceous (figure 7, f–o)

Calpionellids are tiny, bell-shaped objects (loricae are 60–150 microns high) widely distributed in the fine-grained limestones of the uppermost Jurassic and Lower Cretaceous, in the region of Tethys Sea. As their loricae are composed of calcium carbonate, it is unlikely that they would be related to tintinnids.

Because it is necessary to examine them in a very finely ground thin section, it must be remembered that different sections of a single species may appear dissimilar. In the fine-grained limestones that are generally devoid of other fossils, Calpionellidea represent good guide fossils.

2. Subkingdom **Metazoa (multicellular animals)**

The body of the Metazoa is composed of numerous morphologically and physiologically differentiated cells. Within the metazoan organism, various cells serve different functions.

The body of all metazoans is built of layers, of which there are at least two: the outer, protective ectoderm, and the inner, digestive endoderm. Both develop from the early germ layers (primordial tissues).

According to the degree of development between ectoderm and endoderm, either may remain a more or less partitioned primordial cavity, or a mass of cells and intercellular material—the mesenchyme—may develop. In higher Metazoa, the Coelomata, the space between ectoderm and endoderm is taken up by mesoderm. In contrast to the mesenchyme, the mesoderm is a layer of epithelial cells. Its formation differs from group to group. The (secondary) body cavity within the mesoderm is called the coelom.

Three divisions of Metazoa are recognized: **Mesozoa, Parazoa,** and **Eumetazoa.** The very primitively organized Mesozoa do not occur as fossils.

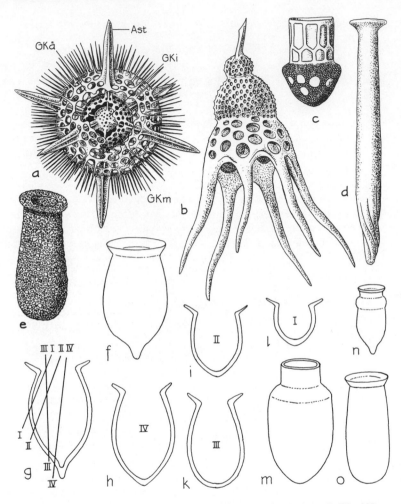

FIGURE 7. a–b Order Radiolaria. **c–e** Order Ciliata, loricae of recent Tintinnidae, 120–450x. **f–o** Calpionellidea. **a** Suborder Spumellina: *Hexacontium* (outer and middle spheres partially broken away), 200x, Recent. **b** Suborder Nasselina: *Cryptoprora*, 150x, Recent. **c** *Dictyocysta*. **d** *Tintinnus*. **e** *Tintinnidium*. **f–l** *Tintinnopsella*, showing different sectional views (I–IV), 300x, Lower Cretaceous. **n** *Crassiolaria*, 200x, Upper Jurassic. **o** *Tintinnopsella*, 200x, Lower Cretaceous. Ast, axial spine, GKa, GKi, GKm, outer, intermediate, and inner lattice sphere (after *Campbell, Haeckel, Kofoid* and *Campbell, Reichenow* and *Brandt, Remane*).

Division **Parazoa**

The parazoan body is an aggregation of mesenchymatic cells. True animal organs, such as the nervous system, sensory organs, or muscles, are not developed. There are a few different types of cells, all exhibiting a high degree of plasticity.

All recent representatives of Parazoa belong to the phylum Porifera (sponges). Archaeocyatha, a fossil group limited to the Cambrian, are extinct parazoans. Porifera cannot be regarded as a connecting link between Protozoa and Eumetazoa. Both Parazoa and Eumetazoa probably evolved from the Protozoa separately. Porifera can be derived from certain flagellates (Choanoflagellata).

1. Phylum Porifera (Spongia, sponges)

Lower Cambrian–Recent (figures 8–11)

Porifera are sessile, predominantly marine animals. The structural cells form epithelial layers only on the surface of the body and along the canals that traverse it. Depending on the point of view, two groups or three layers of the sponge body can be recognized, each characterized by certain types of cells. The cover layer (pinacoderm), which forms the surface of the body and lines the in- and excurrent canals, consists of flat, mostly nonciliate pinacocytes. The chambers inside the body are lined by choanocytes, cells with collared flagella. By the movement of their flagella, the choanocytes maintain a current of water that runs through the sponge body. Between the pinacoderm and the endoderm is the middle layer, designated as mesenchyme or mesogloea. It contains various types of cells capable of amoeboid locomotion within the undifferentiated mass of tissue. The cover and middle layers together can be designated as the dermal group, the endoderm as the gastral group.

The sponge body is attached by its lower part to the substrate. The body wall surrounds a central cavity (paragaster or spongocoel). Water enters through ostia and, via the incurrent canals (epirrhysa) and their ramifications, reaches the ciliated chambers with choanocytes. There nutrients are absorbed, and the water passes through fine collecting canals into the excurrent canals (aporrhysa) and through their openings (postica) into the paragaster, which opens to the exterior through an osculum.

SKELETON (figures 8, 10)

The soft body of the sponges can be supported by a skeleton of spongin fibers, small calcareous elements, calcareous fibers, or continuous calcareous walls. The diversely shaped skeletal elements

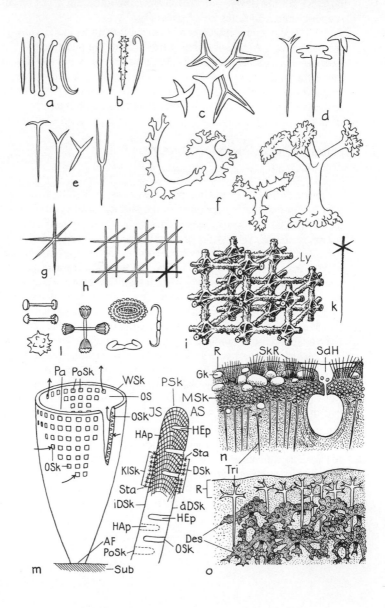

(spicules, sclerites, micro- and megascleres) are composed of opaline silica (amorphous, hydrated silica) or of calcium carbonate.

In addition to the larger (0.1–1 mm) megascleres, many sponges contain microscleres, considerably smaller (0.01–0.1 mm). The sponge skeleton proper consists of megascleres, either loose in the soft body wall, meshed together, or fused on the ends. The many-shaped microscleres are always loose and most frequently occur near the layer boundaries. Classification of fossil sponges is based primarily on the shape and arrangement of the megascleres. Fossil-preserved microscleres are rare. Nearly all scleres have a central axial canal that is (in the living sponge) filled with organic fibers. During diagenesis this canal can become either widened (frequently occurring in siliceous scleres) or narrowed (common in the calcareous scleres).

According to the number of axes along which the scleres are organized, they are classified as monaxon, triaxon, tetraxon, or poly-axon types. In addition, there are desmas, which are irregular. Mon-actinal, diactinal, tetractinal, hexactinal, and polyactinal scleres are distinguished by the number of rays. Desmas may have root-shaped protuberances, with which they could be fairly rigidly connected, albeit not fused. These desmas are designated as tetraclones, rhi-zoclones, or megaclones, according to their form and size.

The form of the skeleton and, consequently, the body form of a sponge is, as a rule, not fixed but rather adapted to the space, sub-strate, and other environmental factors. Common are cup-, funnel-, and dish-shaped sponges as well as those with spherical, cylindrical, and irregularly lobular bodies. In a few species and genera, however, the shape of the body is reasonably fixed and, as such, is of taxonomic importance as well.

FIGURE 8. Porifera. **a–k** Megascleres and lattice types: **a** Monaxon diactinal, **b** Monaxon monactinal, **c** Tetraxon, **d** Triaene, **e** Triradial, **f** Desma, **g** Hexactinal, **h** Hexactinellid dictyid, **i** Hexactinellid lychniskid, **k** Stauract. **l** Microscleres. **m** Structure of a hex-actinellid sponge skeleton. **n** Partial section of *Geodia* (Demospongea). **o** Partial section of a lithistid sponge. äDSk, outer layer of the skeleton, AF, attachment surface, AS, outer side, Des, desma, DSk, dermal skeleton, Gk, stone particle, HAp, cavities of the skeleton for aporhysae, HEp, cavities of the skeleton for epirhysae, iDSk, inner skeletal surface, IS, inner side, KLSk, cloacal skeleton, Ly, lychnisk, MSk, microscleres, OS, osculum, OSk, ostium, Pa, paragaster (spongocoel), PoSk, posticum, PSk, parenchymal skeleton, R, crust of the dermal membrane, SdH, subdermal cavity, SkR, scleres of the crust of the dermal skeleton, Sta, stauract (stab sclerite), Sub, substrate, Tri, triaene, WSk, skeletal wall. Arrows show direction of the water flow (after *Hyman, Moret, Mueller, Rauff, Rieber, Zittel*).

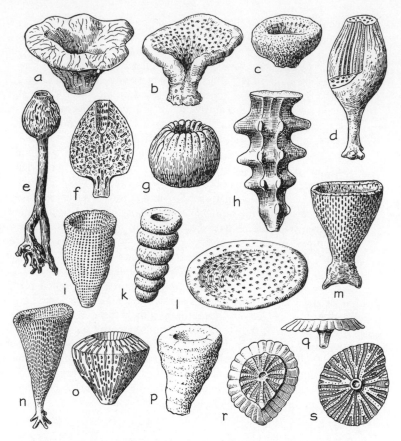

FIGURE 9. Some important Porifera with siliceous skeletons. a–g Class Demospongea: Order Lithistida. h–s Class Hexactinellida: Order Lyssakida (**h**), Order Dictyida (**i–m**), Order Lychniskida (**n–s**). a *Cnemidiastrum*, 0.5x, Upper Jurassic; b *Verruculina*, 0.5x, Upper Cretaceous; c *Callopegma*, 0.3x, Upper Cretaceous; d *Jerea*, 0.3x, Upper Cretaceous; e *Siphonia*, 0.3x, Upper Cretaceous; f *Siphonia*, vertical section, 0.5x, Upper Cretaceous; g *Astylospongia*, 0.5x, Silurian; h *Hydnoceras*, 0.2x, Devonian; i *Craticularia*, 0.5x, Upper Jurassic; k *Caesaria*, 0.5x, Upper Jurassic; l *Stauroderma*, 0.25x, Upper Jurassic; m *Tremadictyon*, 0.3x, Upper Jurassic; n *Ventriculites*, 0.3x, Upper Cretaceous; o *Pachyteichisma*, 0.3x, Upper Jurassic; p *Cypellia*, 0.3x, Upper Jurassic; q–s *Coeloptychium* (side view, top view, bottom view), 0.25x, Upper Cretaceous (after *Rieber, Zittel*).

CLASSIFICATION

The existing classification of the Porifera had to be revised and supplemented in light of recent discoveries of new sponges found by divers in underwater caves, particularly off the coast of Jamaica, and as a result of new information obtained from the Cassian strata (Triassic). Altogether the taxonomy is unconsolidated, since the validity of classification of several groups of sponges is not yet firmly established.

Three classes can be distinguished: Demospongea, Hexactinellida, and Calcispongea. The representatives of Demospongea have no skeleton, a skeleton that consists of spongin with or without siliceous spicules, or an exclusively siliceous skeleton with spicules that are never triaxon. Hexactinellida, or Hyalospongea, have triaxon scleres of opaline silica. Calcispongea, or Calcarea, have triactinal and monaxon scleres of calcite or of calcitic or aragonitic fibers and walls. Secondary siliceous scleres may occur within the calcareous walls as well.

1. Class Demospongea

Cambrian–Recent (figures 8, o, 9, a–g)

Of the eight orders of this class, only Lithistida produced many fossils. Remains of members of the order Keratosida, whose skeleton consists, as in the bath sponge, of easily decomposed spongin, are extremely rare. Monaxon and tetraxon scleres of the other orders are frequently found isolated in rocks. Rocks with a high content of scleres are called spiculite. The boring sponges of the genus *Cliona*, known since the Devonian, play an important role in aiding the disintegration of rocks. By splitting tiny fragments off shells and rocks, they create fine, branching tunnels.

Order Lithistida

Cambrian–Recent (figures 8, o, 9, a–g)

Lithistid skeletons consist of siliceous desmas that may either lie loose in the soft body or be rather firmly fixed by their interlaced processes. The shape and size of the desmas are good taxonomic characteristics.

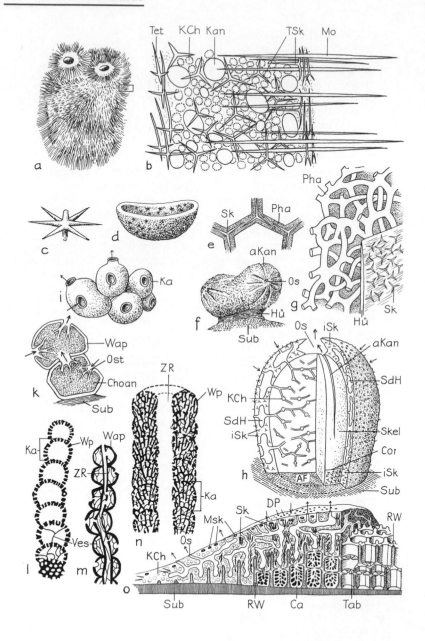

2. Class Hexactinellida (Hyalospongea, glass sponges)

Cambrian–Recent (figures 8, 9, h–s)

The parenchymal skeleton of the Hexactinellida is comprised of triaxon, hexactinal, siliceous sclerites. There are three orders: **Lyssakida, Dictyida,** and **Lychniskida.**

Sclerites of the order **Lyssakida** (Lower Cambrian–Recent) are often only four-rayed, with one of the axes of the triaxon reduced. Scleres are not fused together, which is why complete skeletons of the fossil Lyssakida are rare. Lyssakida were relatively numerous during the Devonian; from Permian and Triassic they are so far unknown. For this reason many authors assume that the Lyssakida that were making a hesitant comeback during the Jurassic evolved independently of the Paleozoic ones. Recent Lyssakida are common in deeper seas.

Triaxons of the order **Dictyida** (Middle Ordovician–Recent) are fused together in a strong, roughly rectangular lattice. Sclerites are noticeably more densely packed in the firmer outer and inner layers of the parenchymal skeleton. The outer layer is perforated by ostia; the inner layer by postica. These are openings leading to the inner skeletal cavities that contain the incurrent and excurrent canals, epi- and aporrhyses. The outer- and innermost parts of the body wall

FIGURE 10. a–n Class Calcispongea. o Class Sclerospongiae. a, b Order Dialytina, *Leuconia,* whole sponge and a section of the wall (much enlarged), Recent; c, d Order Octactinellida, *Astraeospongium,* octactinal spicule (5x) and the entire sponge (0.5x), Silurian–Devonian; e–h Order Pharetronida, suborder Inozoa; e Pharetrone fibers with calcareous scleres, f *Enaulofungia,* 0.7x, Jurassic, g Schematic of the skeletal structure of the Inozoa, h Internal structure of the Inozoa (left, part of the body removed, right, cortex removed); i–n Order Pharetronida, suborder Sphinctozoa, i *Celyphia,* 2.5x, Triassic, k Schematic of a section of two chambers of *Celyphia,* 4x, l *Colospongia,* 3x, Triassic, m *Girtyocoelia,* 4x, Triassic, n *Dictyocoelia,* 6x, Triassic. o *Merlia,* schematically left, with choanosome right, choanosome removed, 20x, Recent. AF, attachment surface, aKan, excurrent canal, Ca, calocytes (amoeboid cells), Choan, choanosome, Cor, cortex, DP, dermal pores, Hü, outer layer, iSk, isolated sclerites, Ka, chamber, Kan, canal, KCh, canal with choanocytes, Mo, monaxon scleres, Msk, microscleres, Os, osculum, Ost, ostium, Pha, pharetronid fiber, RW, tube wall, SdH, subdermal cavity, Sk, scleres, Skel, skeleton, Sub, substrate, Tet, tetraxone scleres, TSk, triradiate scleres, Ves, vesiculae (dissepiments), Wap, aporate wall, Wp, porate wall, ZR, central tube, arrows indicate direction of the waterflow (after *Hyman, Kirkpatrick, Laubenfels, Ott, Rieber, Ziegler, Ziegler* and *Rietschel*).

(dermal and gastral membranes) in live sponges are supported by isolated wall sclerites. These modified hexactines are anchored in the parenchymal skeleton by their long rays and with the other rays support the peripheral parts of the sponge. As the sponge decays, the sclerites usually fall off; they are seldom preserved together with their parenchymal skeleton.

In the order **Lychniskida** (Jurassic–Recent), the cross-joints of the fused hexactine sclerites are lychnisk-shaped. In other characteristics, their skeleton is largely similar to that of the Dictyida.

3. Class Calcispongea (Calcarea, calcareous sponges)

Cambrian–Recent (figure 10)

Sponges with skeletons of calcite or aragonite. When sclerites are present, they are monaxon, triactinal, or tetraxon.

1. Order Dialytina

Carboniferous–Recent (figure 10, e–h)

Skeleton of monaxon, triactinal, and tetraxon sclerites, loosely stuck in the body wall. Most recent Calcarea, which prefer warm shallow waters, belong to this order.

2. Order Pharetronida

Cambrian, Carboniferous–Recent (figure 10, e–h)

The skeleton consists essentially of calcareous fibers or walls. Free sclerites occur only secondarily or are absent. To what extent the original sclerites were lost through the diagenetic processes is subject to different interpretations. Pharetronida are generally small (in the order of 0.5–5 cm in diameter and 1–15 cm long). Most of them inhabit warm, shallow waters where they frequently occur together with reef corals. They were important reef builders primarily during the Permian and Triassic.

Skeletons of the suborder **Inozoa**, or **Pharetronida** sensu stricto (Permian–Recent), are networks of interlaced calcareous (0.1–0.5 mm in diameter) fibers. Triactinal sclerites are incorporated within this network in Jurassic to recent Pharetronida. Sclerites also occur outside the pharetrone fibers, especially in the border layers of the body. In Triassic inozoans the pharetrone fibers are composed of aragonite, but in the less numerous Recent forms they are of calcite. The change probably started as early as the Lower Cretaceous. In addition to the composition of the fibers, an important characteristic is their microstructure.

In the suborder **Sphinctozoa** or **Thalamida** (Cambrian, Carboniferous–Recent), the skeleton is always partitioned and consists of calcareous walls. The chambers may remain hollow, or a calcareous fill may be produced: reticular, tubular, vesicular, or trabecular. The body may enclose a central tube (central cavity) connecting the individual chambers. Not counting the inhalant and exhalant openings, the chamber walls may be either entirely without pores (**Aporata**) or perforated by numerous fine pores (**Porata**). The wall, which is composed partly of calcite and partly of aragonite, exhibits various microstructures (spherolithic, irregular, or clinogonal). Sphinctozoa were most widely distributed during the Permian and Triassic.

3. Order Octactinellida

Lower Cambrian–Lower Carboniferous (figure 10, c, d)

The skeleton of these sponges consists of many-rayed sclerites of uncertain original composition (calcareous or siliceous). The inclusion of this order with Calcarea is therefore not final.

4. Class Sclerospongiae

Ordovician–Recent (figure 10, o)

Sponges with skeletons of calcareous tubules that vary in cross section and diameter (0.1–1.2 mm). Monactinal and some polyactinal sclerites also occur sporadically in the choanosome. Sclerites were found in several fossil representatives even in the walls of the tubules. The tubules are partitioned by horizontal tabulae and may have spiked walls. Choanosome, the living part of the sponge, covers the very massive skeleton like a skin, a few millimeters thick. Many recent as well as fossil forms developed astrorhizae, shallow grooves in the skeleton surface, radiating from common central points. The class Sclerospongiae was established in 1970, mainly on the basis of recent forms discovered on the north coast of Jamaica. Several fossils whose taxonomy was much disputed emerged as belonging to this class. Chaetetidae (Ordovician–Recent), for example, were classified as Sclerospongiae of a new order, **Chaetetida**.

Chaetetida are flat, lumpy, pillow-shaped or finger-shaped, and supported by round or polygonal calcareous tubes with horizontal tabulae. They occur mainly in sediments of shallow, warm waters, sometimes together with corals.

Order Stromatoporoidea

Cambrian–Cretaceous, ? Recent (figure 11)

Stromatoporoidea are treated here as an order of the class Sclerospongiae because their important skeletal characteristics are similar, if

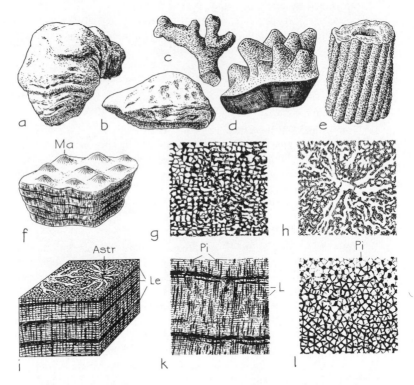

FIGURE 11. Order Stromatoporoidea. **a–e** Growth forms. **f–h** *Stromatopora*, Ordovician–Permian, **f** Part of a specimen, 1x; **g** Longitudinal section, 7x; **h** Upper surface with an astrorhiza, 7x. **i–l** *Actinostroma*, Cambrian–Lower Carboniferous. **i** Part of a specimen, 1.5x, **k** Vertical section, 1.5x, **l** Section parallel to the surface, 6x. Astr, astrorhiza, L, laminae, Le, latilaminae, Ma, mamelon, Pi, pila (after *Lecompte, Moore*).

not identical, to those of both fossil and recent representatives of this class. Earlier they were classified as hydrozoans, and most recently they were connected with stromatoliths. The sponge model, developed for the Stromatoporoidea a few years ago, explains their forms as well as or much better than the former hydrozoan model.

The calcareous skeleton of the Stromatoporoidea consists of horizontal (parallel to the surface) and vertical elements. Horizontal laminae, which account for the layered structure of this group, are supported by vertical pilae. Many skeletons contain latilaminae, which present a unified appearance because of thickening and pigmentation. It is here that the pilae primarily connect. Latilaminae probably reflect periods of slower growth. Several forms have vesicular structures instead of the pilae and laminae, and many show astrorhizae, the radiating depressions. The sponge-model connects their presence with the function of the excurrent canals. On the surface of many forms are small, raised, knoll-like swellings, called mamelons. The usual shapes of the stromatopore skeleton are crusts, massive low domes, or lumps and knobs; less common are stalks and branching shapes. Size varies, from less than 1 cm to more than 1 m. Stromatoporoidea were important reef builders, particularly during the Silurian and Devonian periods.

ECOLOGY OF THE PORIFERA

The versatile Demospongea occur from shallows to depths of several hundred meters. Of these, the extant Lithistida are most often found in warm seas between 100 and 350 m. Recent Hexactinellida live on ooze substrates, from the tidal zone down to the depths of 200–500 m. Most Calcispongea today inhabit shallow warm waters where they frequently occur together with hermatypic corals. Most fossil Pharetronida lived in similar habitats.

During geological history, the sponges have repeatedly emerged as reef builders, in the company of other organisms, primarily stromatolites: Lithistida during the Lower Ordovician and Upper Jurassic, Hexactinellidea in the Upper Jurassic, Inozoa and Sphinctozoa mainly in the Permian and Triassic, Chaetetida during the Carboniferous and Upper Jurassic, and Stromatoporoidea primarily during the Silurian and Devonian. A well-known example is the Upper Jurassic sponge-algal reefs of the Swabian-Franconian mountains (the Swabian-Franconian Alb).

2. Phylum Archaeocyatha

Lower and Mid-Cambrian (figure 12)

Archaeocyatha are a diverse, marine group of animals, with a cal-careous skeleton that somewhat resembles both the sponges and the corals. The overall perforated skeleton is usually goblet-, bowl-, or cup-shaped. The average skeleton size of the most common, sharply conical forms is 10–25 mm in diameter and 80–150 mm long.

Most skeletons have a perforated double wall (a single-layer wall rarely occurs) surrounding a central hollow space. The inside and out-side walls are connected by vertical, radially arranged, perforate partitions—the parieties. The space between the inner and outer walls is called intervallum; the space between two neighboring parieties is designated as interseptum. Horizontal (tabulae) or vesicular compo-nents (dissepiments) may occur in the central cavity as well as in the intervalla of different forms.

FIGURE 12. Restora-tion and cross section of a rep-resentative of the phylum Archaeo-cyatha. AW, outer wall, HS, pedal disk, Is, inter-septum, Iv, inter-vallum, IW, inner wall, Pa, paries, ZH, central cav-ity (after *Vologdin*).

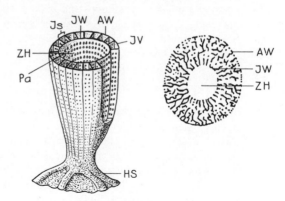

CLASSIFICATION

The phylum Archaeocyatha is divided into classes according to the arrangement of the skeletal structures. Particular attention commands the inner wall, owing to its great complexity and variability. The entire group being extinct since Upper Cambrian, its position among the Parazoa is by no means undisputed.

ECOLOGY AND DISTRIBUTION

Archaeocyatha were fixed sessile benthic forms that inhabited shallow tropical seas. They are found mainly in the calcareous layers of the Lower Cambrian. Their preferred depth of water was 20–50 m; only the smaller forms lived above and below this range, and below 100 m they were absent. The algal-Archaeocyatha reefs were raised only within the 20–50 m depth. Archaeocyatha were widely distributed in Siberia, North America, Australia, and North Africa. Because of their rapid evolution and wide variety of forms, they are widely used in stratigraphic studies, particularly of the Siberian plateau.

Division **Eumetazoa**

Eumetazoa are characterized by the presence of typical animal cells, such as sensory, neural, or muscle cells. Two germ layers (ectoderm and endoderm), from which originate tissues and organs, occur during the ontogeny of the lower eumetazoans (Coelenterata). In the higher Eumetazoa a third layer, the mesoderm, is also formed.

1. Subdivision **Coelenterata**

The radially symmetrical body of adult Coelenterata has a central cavity from which a single opening leads to the outside. The body wall consists of the ecto- and endoderm and an originally noncellular supporting lamella, the mesoglea. In more highly developed Coelenterata the mesoglea may be invaded by ectodermal cells.

Of the two phyla of the Coelenterata, Cnidaria and Acnidaria, only the Cnidaria are well documented as fossils.

Phylum Cnidaria

(figures 13–18)

This phylum includes Coelenterata that are equipped with stinging ectodermal cells (nematocysts, cnida). These complex cells function as protection and for obtaining prey. Cnidaria are either solitary or colonial. Alternation of generations is marked by two distinct forms: the sessile polyp and the free-swimming medusa. In different classes of the Cnidaria, either the polyp or the medusa may be more highly developed, or one or the other may be absent.

Reproduction is both sexual and asexual. Asexual reproduction, which is limited to the polyp stage, leads to the formation of colonies, the new polyps remaining connected to the parent.

Three classes are distinguished by the structure of the polyps and, where developed, by the structure of the medusae: Hydrozoa, Scyphozoa, and Anthozoa.

1. Class Hydrozoa

The tiny polyps of most colonial hydrozoans have one simple, undivided gastrovascular cavity. The polyps' shape reflects their different functions within the colony. The epidermis of many species secretes a chitinous cuticle, the periderm, to the outside. Calcium carbonate (as aragonite) may be deposited within the elastic periderm, especially in the basal parts. Fossil remains of forms with such calcareous periderm are well preserved. *Millepora,* occurring since Upper Cretaceous, is an important reef builder.

Stromatoporoidea, currently usually classified among Calcarea or Sclerospongiae, were formerly placed among the Hydrozoa. True representatives of the Hydrozoa are common only since the Cretaceous, though the oldest ones had been described from the Cambrian.

2. Class Scyphozoa

Marine Coelenterata in which the medusa generation is dominant. The gastrovascular cavity of the polyps is divided by four radial mesenteries into four compartments. The medusae are tetramerous as well.

1. Subclass Scyphomedusae

? Lower Cambrian, Upper Jurassic–Recent

All recent Scyphozoa belong to this subclass. They are very rarely preserved as fossils, because neither the polyps (always small) nor the medusae (jellyfish), which often grow to a diameter of 2 m, possess any hard parts. Most fossil jellyfishes were found in the Solnhofen Lithograhic Limestones (Upper Malm) as imprints. Fossil polyps of the Scyphozoa are unknown.

2. Subclass Conulata

Mid-Cambrian–Lower Triassic (figure 13)

Pointed quadrilateral pyramids, with thin walls of chitin or a chitino-phosphatic substance are considered to be the peridermal shells of the polyp generation of fossil Scyphozoa. These shells are normally 5–12 cm long. The lateral faces are often marked with a pronounced median longitudinal groove as well as by distinct horizontal or less distinct longitudinal ridges and furrows. Many forms have grooves in place of the edges of the pyramid. Sideways extensions of the base may have served as a lid. The most important Conulata are gathered in the suborder Conulariina (Cambrian–Triassic).

Most Conulariina were attached to the substrate by a pedal disk. The base of the pyramid, facing upward, was taken up by the oral disk and the tentacle ring. Attached along the four indented median lines were the mesenteries. Many Conulariina appear to have lived planktonic, in which case the tip of the pyramidal or subcylindrical shells pointed up, and the bases with tentacles pointed down.

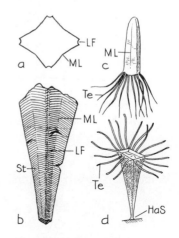

FIGURE 13. Subclass Conulata. **a, b** *Paraconularia,* cross section and side view, 0.4x, Lower Carboniferous; **c** Restoration of planktonic *Exoconularia,* 0.2x, Ordovician; **d** Restoration of a sessile conularian, 0.5x, Devonian. HaS, pedal disk, LF, longitudinal furrow, ML, median line, St, cross-furrows, Te, tentacles (after *Kiderlen, Moore* and *Harrington*).

3. Class Anthozoa

(figures 14–18)

Marine—solitary or colonial—Coelenterata in which only the sessile polyp generation is developed. The polyps have a gullet extending from the mouth into the enteron. The enteron is divided by vertical, radially arranged mesenteries into numerous compartments. The proximal end of each mesentery is swollen or thickened. The endodermal cells form a long, thickened muscle on one side of the mesentery. These muscles aid the polyp's orientation. The class Anthozoa includes three subclasses: Ceriantipatharia, Octocorallia, and Zoantharia; only Zoantharia yielded many fossils.

Subclass Octocorallia

? Precambrian, Permian–Recent

Small polyps of the colonial Octocorallia have eight feathered tentacles, and their enteron is divided by eight simple mesenteries into eight compartments. New polyps of a colony evolve from fine endodermal tubes (solenia) extending from the gastrovascular cavity into the cellular mesoglea. In the mesoglea epidermal scleroblasts produce sclerites (= spiculae) of either calcite or keratinous scleroprotein gorgonin. Calcareous sclerites, 0.01–0.5 mm long, are often gnarled, spindle-shaped structures; their form is species- and genus specific.

The polyps sit in small tubes into which they can retreat. Many Octocorallia also form solid fixed skeletal structures, keratinous or calcareous.

The Australian **Ediacara-Fauna** contains imprints of organisms similar to the recent Pennatulacea (one order of Octocorallia). However, it is questionable whether the fossil *Rangea* is indeed an octocorallian.

Most extant species of Octocorallia live on the continental shelf, avoiding, however, the tidal zone. They occur in cold as well as warm waters and have been found as deep as 4,400 m.

Subclass Zoantharia (stony corals)

Cambrian–Recent

Zoantharian polyps carry numerous nonciliated tentacles; their mesenteries are always paired. Zoantharia occur both alone and in colonies. Seven suborders are recognized, distinguished by the arrangement of the mesentery pairs and when possible by the skeletal structure. Only those orders that form calcareous skeletons are fossil documented, namely, Rugosa, Heterocorallia, Scleractinia, and Tabulata.

SKELETON (figures 14 and 15)

Calcareous external skeletons of the Zoantharia are secreted by the ectoderm. The skeleton of an individual polyp is termed corallite; that of an entire colony of polyps is called the corallum. Between neighboring corallites of a corallum may be formed calcareous connective material (coenosteum or sclerenchyme). The formation of a skeleton begins with a basal plate, secreted by the pedal disk of a polyp. The lateral extension of the basal plate is termed epitheca. When it is developed, it surrounds the corallite as an outer wall, frequently showing growth lines. Septa are radially arranged vertical walls originating in pouches of the soft body. Endosepta and exosepta are distinguished in Scleractinia. Endosepta arise in the endocoeles (space between mesenteries of the same pair), exosepta in the exocoeles (spaces between two adjacent pairs of mesenteries). The first six septa are called protosepta, and all the following are called metasepta. Distal parts of the septa may be united by the outside wall (the theca) or by other similar structures. Axial structures such as the columella frequently occur in the center of a corallite. Pali or paliform lobes are sometimes found at the inner edges of the septa. During their growth many forms secrete calcareous substances that separate living parts from the abandoned, dead ones along the periphery or inside the corallite. Horizontal skeletal elements in the central portion are designated individually as tabulae, collectively as tabularium. Vesicular structures appearing mainly along the periphery and bracing the septa are named dissepiments, their entire system the dissepimentarium. In many forms the build-up of septa is preceded by the formation of vesicular praesepiments, located on the inside of the wall of the corallite. The septa formed subsequently are attached to the praesepiments. Many members of the Rugosa exhibit cardinal or alar fossulae (depressions within a limited area of the corallite).

Skeletal Structure (figure 15)

The skeleton of the Scleractinia consists primarily of aragonite. Still under discussion is the question of whether the skeletons of the Paleozoic Zoantharia (often excellently preserved) were also of aragonite or whether they consisted of calcite. Skeletal units, which are formed in folds of the body wall, have a trabecular structure. Trabeculae consist of linearly arranged sclerodermites: bundles of fine calcareous filaments, radially arranged around calcification centers. The septa, the theca proper, and the columella all show a trabecular structure. All skeletal elements that are secreted to one side (not in folds) by the

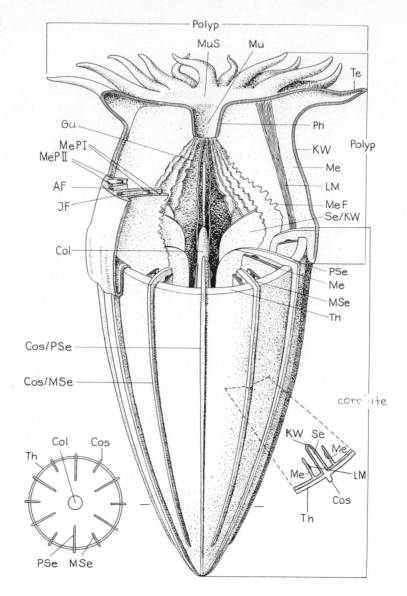

FIGURE 14. Schematic drawing of a solitary representative of the order Scleractinia, showing various sectional views. AF, exocoele (space between adjacent pairs of mesenteries), Col, columella, Cos, costa, Cos/PSe, costa of the primary septum, Cos/MSe, costa of the metaseptum, IF, entocoele (space within pairs of mesenteries), KW, body wall, LM, longitudinal muscle, Me, mesentery, MeF, mesenterial filament, MeP, I–II mesentery pair of the first and second order, MSe, metaseptum, Mu, mouth MuS, oral disk, Ph, pharynx/gullet, PSe, primary septum, Se, septum, Se/KW, septum encased in the body wall, Te, tentacle, Th, theca.

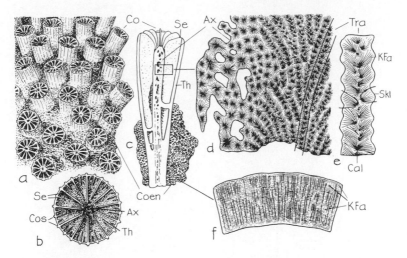

FIGURE 15. Skeletal structure of *Galaxea* (order Scleractinia). **a** Part of the corallum, 1x; **b** Corallite, top view; **c** Corallite, longitudinal section; **d** Wall-parallel section of a septum (right), and cross section of the axial structure (left); **e** Septum in a vertical section, 25x; **f** Part of a dissepiment of the coenosteum, 25x. Ax, axial structure, Cal, calcification centers, Co, corallite, Coen, coenosteum composed of dissepiments, Cos, costae, KFa, calcareous fibers, Se, septa, Skl, sclerodermite, Tra, trabecula, Th, theca (after *Wells*).

ectoderm (i.e., basal plate, epitheca, tabulae, and dissepiments) consist of compact layers in which the calcareous fibers lay more or less parallel to the surface. A secondary layer of calcareous reinforcing filaments that may occur anywhere is designated as stereome.

Ontogenic Development of the Septal Structures (figure 16)

Although septa of the typical Rugosa are laid one after the other (serial build-up), typical Scleractinia form septa simultaneously, in circles (cyclic build-up).

The serial build-up and the fact that metasepta are formed only in four of the six existing sectors make the Rugosa bilaterally symmetrical. On one side of the corallite the two septa on both sides of the cardinal septum diverge outward, while on the opposite side they run generally parallel to the counterseptum. Scleractinia normally form six septa in the first and second cycles, twelve in the third, twenty-four in the fourth cycle, and so on. During further growth, exosepta may split in two on the outside; later, endosepta are formed

38

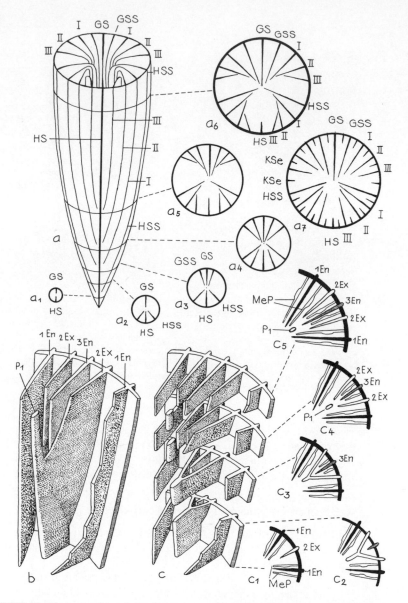

FIGURE 16. Diagrams of septal development in Zoantharia. **a–a₇** Order Rugosa. **b, c–c₅** Order Scleractinia. **a** Corallite from the cardinal side. **a₁–a₇** Cross-sectional views of different ontogenetic stages. **b** Sector of a corallite. **c–c₅** Sector of a corallite dissected, with indication of the mesentery pairs.

En, endoseptum, Ex, exoseptum, GS, counterseptum, GSS, counterlateral septum, HS, cardinal septum, HSS, alar septum, KSe, minor septa, MeP, mesentery pair, P, palus, 1–4 and I–III rank according to the ontogenetic succession (after *Wells*).

in their original locations (figure 16, $c-c_5$). The inner or central part of such a split and substituted septum remains in place as an isolated palus.

CLASSIFICATION

The main classification is based on the ontogeny of the septal structures and on the form of the skeletal elements of corallites and coralla.

1. Order Rugosa

Cambrian–Permian (figures 16, a_1-a_7, 17, a–k)

Solitary and colonial Zoantharia, normally with serial build-up of the septa. Epitheca, when present, is wrinkled (rugose). Many representatives have tabulae and/or dissepiments in addition to the septa. Numerous solitary ones have a cardinal fossula, or alar fossulae. Many of the Rugosa were important reef builders, especially during the Silurian and Devonian periods.

2. Order Scleractinia

Triassic–Recent (figures 16, b–c, 17, l–p)

Solitary and colonial Zoantharia with cyclical build-up of the septa. There are many diverse forms, with corallites from approximately 1 mm to more than 15 cm in diameter. The coralla may attain a size of several meters.
    Among the Scleractinia are distinguished **hermatypic** or reef corals and **ahermatypic** corals. Because of their symbiosis with the monocellular algae (Zooxanthella), the hermatypic corals live only in water penetrated by light, optimally less than 20 m deep, with temperatures between 25 and 29°C and salt concentrations between 34 and 36 promille. In addition, a firm substrate is required for proper growth. Hermatypic Scleractinia have been the most important reef builders since the Mesozoic. Ahermatypic Scleractinia, both solitary and colonial, occur in depths up to 6,000 m, even in cold water.

3. Order Heterocorallia

Lower Carboniferous

This order contains only two genera. The four protosepta are peripherally split into several branches.

4. Order Tabulata

Mid-Ordovician–Permian (figure 18)

This order encompasses exclusively colonial forms. Thin, tubular corallites either adjoin directly or are separated by coenosteum. They always have tabulae, but the septa may be poorly formed, often only as rows of spikes or thorns, or be absent. Corallites may be connected by pores. It is uncertain whether all forms currently classified as Tabulata actually belong to Zoantharia. For *Favosites,* for example, a sponge-model was proposed, and *Heliolites* substantially differs from other Tabulata by the presence of coenostea. Tabulata were important reef builders during the Silurian and Devonian periods.

2. Subdivision **Bilateria**

Bilaterally symmetrical Eumetazoa. During locomotion the mouth end of the body is directed forward; consequently, the sensory and nervous organs concentrate at that end, eventually leading to the development of head and brain (cephalization). In addition to the ecto- and endoderm, a third germ layer, the mesoderm, evolves. The intestine usually has a second opening.

FIGURE 17. Important representatives of the subclass Zoantharia. **a–k** Order Rugosa. **l–p** Order Scleractinia. **a** *Streptelasma,* 0.5x, Middle Ordovician–Middle Silurian, a^1–a^3 Cross and longitudinal sections of the *Streptelasma;* **b, c** *Thamnophyllum,* Devonian, median longitudinal section, 1.3x (**b**), Cross section 1.3x, (**b**^{1}), **c** Part of the corallum, 0.4x; **d** *Hapsiphyllum,* 0.8x, Lower Carboniferous; **e, e**^{1} *Holophragma,* 1.3x, Silurian; **f** *Hexagonaria,* 0.3x, Devonian; **g** *Ketophyllum,* 0.3x, Silurian; **h–h**^{1} *Londsdaleia,* longitudinal and cross section of a corallite, 0.8x, Carboniferous; **i** *Cystiphyllum,* 0.5x, Silurian, **i**^{1} Longitudinal section, **i**^{2} Cross section; **k** *Calceola,* 0.5x, Middle Devonian, **k**^{1} Lid from above, 0.5x, **k**^{2} Restoration, longitudinal section; **l–l**^{1} *Montlivaltia,* 0.5x, Triassic–Cretaceous; **m** *Thecosmilia,* 0.7x, Triassic–Cretaceous; **n–n**^{1} *Turbinolia,* side and top view, 4x, Eocene; **o** *Isastrea,* 0.3x (**o**) and 1x (**o**^{1}), Jurassic–Cretaceous; **p** *Thamnasteria,* 0.3x (**p**) and 1.5x (**p**^{1}), Triassic–Cretaceous. AnL, growth lines, CaFo, cardinal fossula, Col, columella, De, operculum, Dis, dissepiment, Epi, epitheca, GS, counterseptum, HS, cardinal septum, Prä, praesepiments, Se, septum, SeD, septal spines, Tab, tabula, Te, tentacle, Wu, root (after *Hill, Wells*).

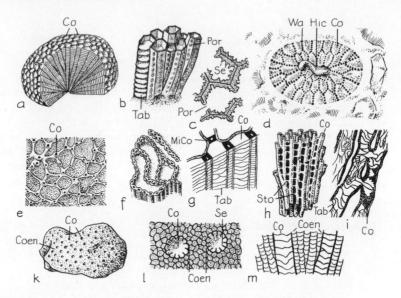

FIGURE 18. Order Tabulata. **a–c** *Favosites,* Ordovician–Devonian, **a** Corallum, partial section, 0.5x; **b** Several corallites, 1.0x; **c** Cross section of two corallites, 2.5x. **d** *Pleurodictyum,* internal mold with a *Hicetes* tube, 0.5x, Lower Devonian. **e** *Aulopora,* 0.5x, Devonian. **f, g** *Halysites,* Ordovician–Silurian, **f** Part of a corallum, 0.5x; **g** Longitudinal and cross section of neighboring corallites, 3x. **h, i** *Syringopora,* Silu-rian–Carboniferous, **h** Part of a corallum, 0.4x; **i** Longitudinal section of neighboring corallites, 2.5x. **k–m** *Heliolites,* Silurian–Devonian, **k** Corallum, 0.5x; **l** Surface of part of the corallum, 5x; **m** Section vertical to the surface, 5x. Co, corallite, Coen, coenosteum, Hic, *Hicetes,* MiCo, microcorallite, Por, pore, Se, septum, Sto, stolo, Tab, tabula, Wa, wall (after *Hill* and *Stumm*).

Phylum Tentaculata

(figures 19–24)

Aquatic, predominantly marine, sessile Bilateria whose oral opening is surrounded by a ring of ciliated tentacles.

1. Class Phoronida

Worm-shaped Tentaculata that secrete a transparent chitin tube, partially encrusted with foreign particles. Some forms make burrows and

bore holes in shells of mussels by chemical action. Scolithus-
sandstone of the Baltic-Scandinavian Lower Cambrian probably
contains tubes of Phoronida. Phoronid traces found in rostra of
belemnites were once described under the name *Talpina*.

2. Class Bryozoa

Ordovician–Recent (figures 19–21)

A large group of exclusively colonial Tentaculata, rich in various
forms. Their soft frontal part (polypide), together with folded-up
crowns of tentacles (lophophore), can retract into a voluminous rear
part (cystid).

Polypide and cystid together form a zooid, which is usually
smaller than 1 mm. Epidermis of the cystid secretes an organic cuticle of protein with carbohydrates. Calcium carbonate may be
embedded in it (mostly as high-magnesium calcite). The calcareous
cuticular part of a zooid is designated as zooecium. A community of
the zooids or the zooecia is a zoarium. Zoaria may be encrusted and
occur in the shape of mats, crusts, lobes, lumps, trees, or funnels, up
to several centimeters in size. There is a division of functions among
the zooids in a zoarium, and the zooids are differentiated accordingly.
Most common are the autozooids, which provide the nourishment and
reproductive functions. The narrow opening where a polypide enters a
cystid is the aperture. In many forms the aperture may be closed by a
lid (operculum) when the polypide retracts.

Zooids with rigid, dense zooecia compensate for the change in
volume caused by protrusion and retraction of the polypide by either
a flexible frontal membrane or by a compensation sac (ascus) if the
front, too, is rigid. The ascus is a pouchlike ectodermal invagination
that may be enlarged by the tension of parietal muscles, thus increasing the pressure inside the cystid and forcing the polypide out.
Cystids of a zoarium are connected together by pores, poral plates, or
missing parts of the dividing walls. Unlike autozooids, the hetero-
zooids serve to anchor (kenozooids) and clean the colony (avicularia
and vibracula); they are also used partially for reproduction (gonozooids). Their zooecia (heterozooecia) are adapted to these functions
and differ from the common structure of the autozooecia accordingly.

In contrast to gonozooids, which are self-contained zooids with a
more or less reduced polypide, the ovicells (ooecia) are but extrusions
or invaginations of the autozooids. They contain fertilized eggs and
are dedicated exclusively to the brood care.

44

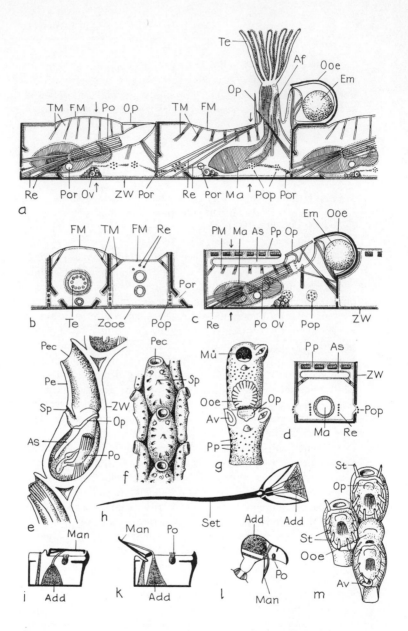

FIGURE 19. Class Bryozoa, order Cheilostomata. **a–m** Schematics of structure. **a, b** Longitudinal and cross section of zooecia of Anasca; **c, d** Longitudinal and cross section of zooecia of Ascophora; **e, f** *Tessaradoma*, longitudinal section, 100x, and part of a zoarium, 50x; **g** *Schizoporella* (two zooecia), 30x; **h** Vibraculum; **i–k** Avicularium without pedicle (closed and open); **l** Avicularium with pedicle, *Bugula*, 70x; **m** *Callopora*, 40x.

Bryozoa reproduce both sexually and asexually. A larva hatches from a fertilized egg of marine bryozoans; it undergoes metamorphosis into a first cystid, which in turn sprouts the associated polypid. This primary zooid (also called ancestrula) produces one or more further zooids by budding (asexual phase). In this manner a bryozoan colony—the zoarium—is created.

CLASSIFICATION (tables 3, 4)

Because only retracted or dead bryozoans are generally available for study, their taxonomy is built primarily around the cystid or the zooecium. The taxonomy is not yet consolidated; researchers disagree in particular about the scope of individual subclasses and the number of orders.

Three subclasses have been distinguished, based on the presence or absence of the epistome and on the structure of cystids and lophophores (table 3).

Abbreviations used in figures 19–21

AcP	acanthopore	MP	mesopore
Add	adductor muscle	Mü	aperture
Af	anus	mZ	mature zone
Anc	ancestrula	Ooe	ooecium
AnF	attachment surface	Op	operculum
As	ascus	Ov	ovary
At	atrium	Pe	peristomium
AuP	autopore	Pec	peristomicium
Av	avicularium	PM	parietal muscle
AZ	autozooecium	Po	polypid
BaL	basal lamina	Pop	pore plate
Ca	carina (median keel)	Por	pore
Coen	coenosteum	Pp	pseudopore
Cy	cystiphragm	RAcP	rib with acanthopores
D	diaphragm	Re	retractor muscle
Dil	dilatator muscle	Set	seta
Dis	dissepiment	Sp	spiramen
Do	spine	St	prickle
Em	embryo	Te	tentacle
FM	frontal membrane	TeS	tentacle sheath
HSe	hemiseptum	TM	transverse muscle
imZ	immature zone	Vo	vestibule
Ma	stomach	Zooe	zooecium
Man	mandible	ZW	zooecian wall
ML	mesotheca	(after *Bassler, Moore, Ryland*).	
Mo	monteculli		

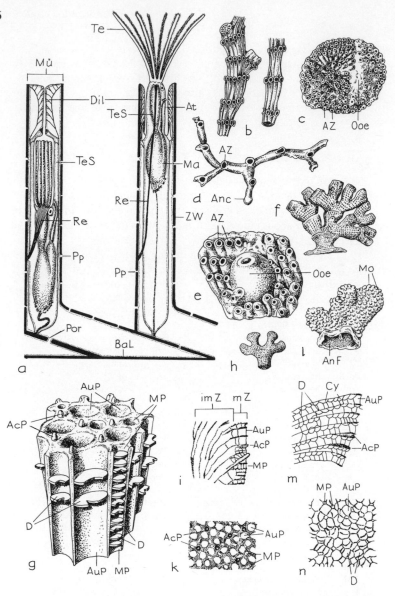

FIGURE 20. Class Bryozoa. **a–f** Order Cyclostomata. **g–n** Order Trepostomata. **a** Schematic of the structure; **b** *Spiropora,* 7x, Jurassic–Recent; **c** *Plagioecia,* 8x, Jurassic–Cretaceous; **d** *Stomatopora,* 9x, Ordovician–Recent; **e, f** *Multisparsa,* Jurassic; **e** Part of a zoarium, 15x; **f** Zoarium, 0.8x; **g** Structure of a zoarium; **h–k** *Dekayella,* Ordovician; **h** Zoarium, 0.3x; **i** Longitudinal section, 10x; **k** Section of the surface, 10x; **l–m** *Monticullipora,* Ordovician; **l** Zoarium, 0.5x; **m** Longitudinal section, 12x; **n** Surface section, 12x. Abbreviations as in figure **19.**

TABLE 3.   Overview of the Subclass Bryozoa

Subclass Geological Range	Epistome	Lophophore	Cystid	Habitat
Phylactolaemata (freshwater bryozoans) Cretaceous–Recent	Present	Horseshoe-shaped, with with 30–100 tentacles	Mostly tubular, not mineralized	Freshwater and brackish water
Stenolaemata Ordovician–Recent	Absent	Circular, with 12–20 tentacles	Tubular or shoe-shaped, calcareous wall	Marine
Gymnolaemata Ordovician–Recent	Absent	Circular, with few tentacles	Usually short, box-shaped, with an operculum, wall often strongly calcareous	Marine

ECOLOGY

Most bryozoans lead sessile, benthic lives in the sea; they are found as deep as 8,300 m. The greatest diversity of species and individuals is found on the continental shelf, optimal conditions being found in 20–80 m depths. Many species are stenohaline and stenobathic. Encrusted zoaria and zooids with shorter lophophores are generally found in shallow and turbulent waters; delicate, nonincrusted zoaria whose zooids often have long lophophores are usually confined to still, deep waters. In many species the growth and shape are affected by the environmental factors. Numerous Bryozoa are specialized to the extent of living exclusively on certain substrates; when the proper substrate is present, so is the respective bryozoan. Many rocks, especially from Upper Cretaceous and Early Tertiary, consist largely of bryozoan remains.

3. Class Brachiopoda

Cambrian–Recent (figures 22–24)

Exclusively marine, solitary, bilaterally symmetrical Tentaculata, normally attached to substrate by a pedicle. The soft body is pro-

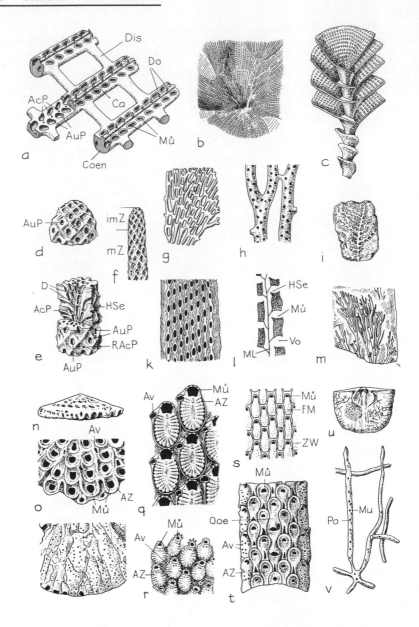

tected and supported by a shell with two valves: the pedicle (ventral) valve and the brachial (dorsal) valve. Concentric growth lines are arranged around the first-formed part, the beak. The body occupies only the posterior end of the mantle cavity; the free anterior part of the space contains tentacle-bearing lophophores, which join the body near the mouth. When the valves are open, coordinated movement of cilia located along the tentacles causes a sideways influx of water. Food particles are filtered out by the tentacles and directed toward the mouth via branchial grooves in the lophophores. The water is exhaled through the front of the shell.

Lophophores of many brachiopods are proximally supported by a calcareous loop, the brachidium. Distally they are usually coiled spirally. In the "hinged" brachiopods (articulates), the pedicle valve is equipped with two teeth that fit into corresponding dental sockets of the brachial valve. The opening muscles (divaricators or diductors) are attached to a processus of the brachial valve behind the hinge and to corresponding attachment areas of the pedicle valve in front of the hinge. The antagonistic closing muscles (adductors) are attached in front of the hinge on both valves. The shell can move on its pedicle by the action of the pedicle muscles (adjustors), which are attached in the beak area of the pedicle valve.

EXTERNAL MORPHOLOGY OF THE SHELL (figure 22)

Shells of most brachiopods are between 1 and 5 cm long or wide. Both valves are usually domed. However, in addition to the biconvex shells, there are also those with a flat or concave brachial valve or,

FIGURE 21. Class Bryozoa. a–m Order Cryptostomata. n–t Order Cheilostomata. u, v Order Ctenostomata. a *Fenestrellina*, part of a zoarium (schematic), 12x; b *Fenestella*, zoarium, 0.5x, Ordovician–Permian; c *Archimedes*, Zoarium, 0.5x, Carboniferous–Permian; d–f *Rhombopora*, Devonian–Permian, d Immature zone, 20x; e Mature zone, 20x; f Part of a zoarium, 5x; g, h *Fenestrellina*, Carboniferous–Permian, g Part of the zoarium, 0.5x; h Part of the zoarium, 5x; i *Acanthocladia*, zoarium, 0.7x, Upper Carboniferous–Permian; k, l *Graptodictya*, Ordovician, k Part of a zoarium, 8x; l Longitudinal section of a zoarium, 12x; m *Pseudohornera*, zoarium, 0.5x, Ordovician–Silurian; n–p *Lunulites*, Cretaceous, n Zoarium, side view, 5x; o Part of the zoarium, top view, 12x; p Part of the zoarium from below, 12x; q *Rhinipora*, part of a zoarium, 12x, Cretaceous; r *Cribrilaria*, part of zoarium, 12x, Eocene–Recent; s *Membranipora*, part of zoarium, 12x, Miocene–Recent; t *Coscinopleura*, part of a zoarium, 12x, Cretaceous; u, v *Vinella*, Ordovician–Recent, u Etched furrows on a brachiopode valve, 0.5x; v Zoarium, 12x. Abbreviations as in figure 19.

TABLE 4. Orders of Stenolaemata and Gymnolaemata with Fossil Representatives

Order (subclass) Geological Range	Zooecia	Form of the Zoarium; Special Characteristics	Period of Main Evolution; Number of Fossil Genera	Figures
Cyclostomata[1] (Stenolaemata) Ordovician– Recent	Mostly tubular; only a few forms with diaphragms, wall porate.	Crusts, dendritic; gonozooecia and ooecia	Cretaceous $\approx$270	20, a–f
Trepostomata (Stenolaemata) Ordovician– Perm, ? Trias	Tubular auto-pores and mesopores with many diaphragms as well as tiny acanthopores	Crusty, lumpy, dendritic; immature and mature regions of the zoarium, walls of neigh-boring zooecia full or divided	Ordovician $\approx$110	20, g–n
Cryptostomata (Stenolaemata) Ordovician– Recent	Mostly short, shoe-shaped, square or hexagonal autopores, mesopores rare, acanthopores numerous	Netlike, lobu-lar, funnel- or tree-shaped; autopores fre-quently with a hemiseptum	Devonian– Permian $\approx$125	21, a–m
Cheilostomata (Gymnolaemata) Middle Jurassic, Creataceous– Recent	Short, shoe-, jug-, or box-shaped with a reinforced aperture and operculum, some with a vestibule; body wall often heavily calcareous	Mat-, disk-, lump-, band-, or tree-shaped; vibracula and avicularia present; forms both with (Ascophora) and without (Anasca) an ascus	Upper Cre-taceous and Tertiary $\approx$390	19 21, n–t
Ctenostomata (Gymnolaemata) Ordovician– Recent	Compressed or tubular; body wall not calcareous	Incrusting and mat-shaped; many forms etch them-selves into the substrate	$\approx$20	21, u–v

[1]Paleozoic representatives (about 50 genera) that have thin-walled irregular tubes with cystopores between zooecia were recently grouped into a separate order, the Cystoporata.

more rarely, the pedicle valve has this shape. A shell that was originally biconvex may, in the course of ontogenic development, become
planiconvex or concavo-convex. The anterior part of the commissure
(the line where the two valves touch) is designated as the frontal
margin, and the posterior part is designated either as the hinge line
(when straight) or as the cardinal margin (when curved). On the posterior end of each valve, between the beak and the commissure, the
articulates have a roughly triangular flat area. When this area occurs
in connection with the hinge line, it is called interarea (dorsoventrally
curved) or planarea (not curved). It is called a palintrope when it is
delimited by the cardinal margin. In the middle of the interarea, planarea, or palintrope is a triangular opening—the delthyrium. It allows
passage of the opener muscles and, in many brachiopods, of the pedicle as well. The delthyrium may be partially or entirely closed by two
calcareous plates (deltidial plates). The structure of this closure, the
so called deltidium, is very important taxonomically. There may be
another triangular opening in the brachial valve—the notothyrium. In
the course of ontogenic development, it is closed by the chilidium.
The pedicle of adult articulate brachiopodes enters either through the
deltidium or through a separate opening (foramen) in the beak area of
the pedicle valve. In some inarticulates, grooves in the posterior margins of both valves form a passageway for the pedicle. The frontal
commissure may be straight or variously folded. A median sinus (the
sulcus) is commonly found in the pedicle valve; there is a corresponding bulge (the fold) in the brachial valve. These two structures
cause the characteristic trilobation of many brachiopode shells.

INTERNAL MORPHOLOGY OF THE VALVES (figures 22, 23)

The most important features of the inside of the shell are the following: articulation, cardinalia, brachidium, median septa or ridges,
and muscular and vascular impressions.

The **articulation** (hinge) normally consists of two teeth in the
pedicle valve and two corresponding sockets in the brachial valve.
The teeth may be propped against the beak area by dental plates (dental lamellae). If the lamellae connect in the median line, they form a
spoon-shaped structure, the spondylium. The spondylium may be
attached to a median septum as well.

Cardinalia is a designation for the complex of structures at the
hinge line or cardinal margin. They consist of the hinge processus,
the hinge plates, dental sockets, brachiophores, crura, and crural
plates. Brachiophores of Orthida are platelike structures serving prob-

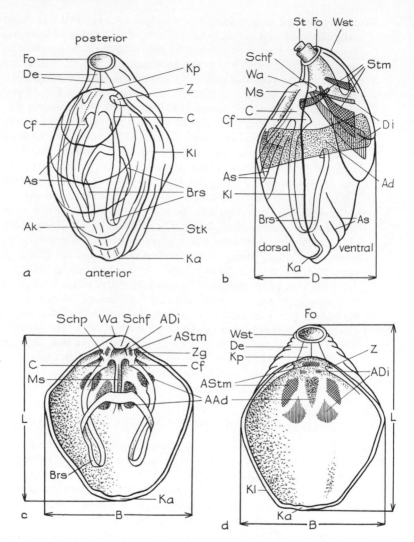

FIGURE 22. Class Brachiopoda, subclass Articulata. Morphology of the shell: **a, b** Entire shell, drawn partially transparent to show the brachidium; **c** Inside of the brachial valve; **d** Inside of the pedicle valve.

ably for lophophore attachment. Crura of Rhynchonellida, Terebratulida, and Spiriferida correspond somewhat to the brachiophores. In Terebratulida and Spiriferida, calcareous support of the lophophore (**brachidium**) is attached to the crura. In the Terebratulida it consists of a short or longer calcareous band that may attach to the median septum; in the Spiriferida it consists of two conical spirals that point laterally or dorsally.

The internal morphology of the shell very often has to be studied in thin section, even in series of sections, because the thin calcareous shells can rarely be separated from the rest of the rock.

COMPOSITION OF THE SHELL

The shell is secreted by the epidermis of the mantle. In most inarticulates the shell underneath the periostracum consists of organic substances (chitin and scleroprotein) with embedded calcium phosphate. The calcium phosphate may be either deposited in thin layers alternating with the organic substance layers (e.g., *Lingula*) or evenly distributed (e.g., *Discinisca*).

Abbreviations used in figures 22–24

AAd	adductor muscle attachment areas (muscle scars)	Kp	posterior commissure
Ad	adductors	L	length
ADi	diductor muscle attachment area	Ms	median septum (median ridge)
Ak	brachial valve	MuA	muscle scar
As	growth lines	PA	planarea
AStm	pedicle muscle attachment area	Pal	palintrope
B	width	Schf	hinge process
Bph	brachiophore	Schp	hinge plate
Brs	loop of the brachidium	Si	sulcus
C	crus	Sp	spondylium
Cf	crural processes	Spi	calcareous spiral
D	thickness	St	pedicle
De	deltidial plates	StK	pedicle valve
Delt	delthyrium	Stm	pedicle muscle
Di	diductors	Stö	pedicle opening
Fo	foramen	W	beak
IA	interarea	Wa	beak of the brachial valve
J	jugum	Wst	beak of the pedicle valve
Ka	anterior commissure	Wu	fold
Kl	lateral commissure	Z	teeth
		Zg	dental socket

(after *Rieber, Zittel*).

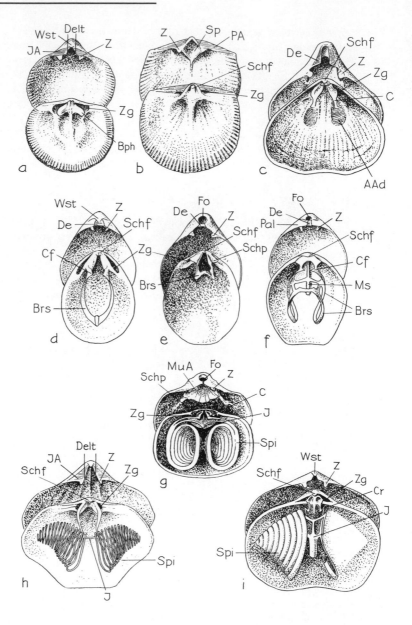

Underneath the periostracum, the shell of articulates consists of calcite. Frequently a laminated structure can be discerned: a thin outer layer of calcite fibers parallel to the surface and a thicker inner layer of fibers that are obliquely arranged. The punctate types of shells develop pores (puncta) that widen toward the outside without, however, penetrating the outer layer. Pores of the living animal contain secretory cells and are connected with the periostracum by extremely fine canals. Impunctate shells have no puncta. Pseudopunctate shells are characterized by numerous obliquely arranged calcite spicules (taleolae). Because the taleolae weather at a different rate than the rest of the shell, eroded shells may falsely appear to be punctate.

CLASSIFICATION (tables 5, 6)

Brachiopoda are usually divided into two subclasses, Inarticulata and Articulata. Most representatives of the Inarticulata have shells of organic matter with embedded calcium phosphate. Only in Craniacea is the shell calcareous (calcite), with a complex system of canals. All Articulata have calcitic shells. Within the Articulata the main taxonomic characteristics used to distinguish individual orders are: composition and form of the shell; presence or absence of brachidium; when present, the structure of the brachidium; and the shape of cardinalia, hinge, deltidia, muscle attachment areas, and vessel impressions.

ECOLOGY

Brachiopoda are benthic organisms, attached to the substrate either by a pedicle or directly by the ventral valve. They feed on organic particles suspended in the water. Most reside on hard substrates, but some have the ability to become anchored by their pedicles to soft

FIGURE 23. Class Brachiopoda, subclass Articulata. Interior morphology of the shell and the brachidia in the orders Orthida (a, b), Rhynchonellida (c), Terebratulida (d–f), and Spiriferida (g–l). a *Resserella*, form with brachiophores, 1.5x, Silurian; b *Vellamo*, 1.3x, Ordovician; c *Hemithyris*, ancistropegmate, 1.5x, Recent; d *Centronella*, ancylopegmate centronellid, 1.0x, Devonian; e *Juralina*, ancylopegmate terebratulid, 0.8x, Upper Jurassic; f *Laqueus* ancylopegmate terebratellid, 0.6x, Recent; g *Gruenewaldtia*, helicopegmate atrypid, 1.0x, Middle Devonian; h *Spiriferina*, helicopegmate spiriferid, 1.0x, Lower Jurassic; i *Athyris*, helicopegmate athyrid, 0.8x, Lower Carboniferous. Abbreviations as in figure 22 (after *Bachmayer, Moore*).

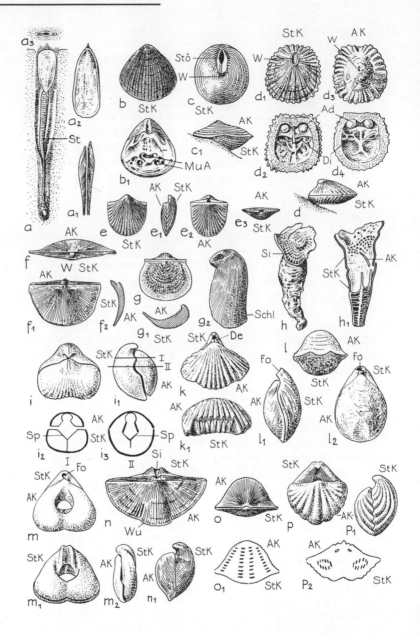

TABLE 5.   Overview of Orders of the Inarticulata

Order Geological Range	Pedicle	Position of the Pedicle Opening	Valve Shape	Shell Composition	Figure
Lingulida Cambrian– Recent	Long and muscular	In the plane of the commissure	Elongated to round	Organic substance + calcium phosphate	24, a, b
Acrotretida Cambrian– Recent	Short or absent	Limited to the pedicle valve or absent	Roundish, flat to sharply conical	Organic substance + calcium phosphate or calcite	24, c, d

bodies and other substrates (e.g., algal stalks). *Lingula* lives in tubular mud burrows.

Shells of recent Articulata are either propped up on the pedicle (pedicle short and thick, foramen wide) or just attached by it (pedicle thin, foramen small). The pedicles of many fossil brachiopods were atrophied, so that these forms, when adult, lay loose in the sediment. Brachiopods adapted to the varied benthic conditions by developing a variety of forms. As a result, they serve as useful facies indicators. Recent brachiopods live predominantly in cool waters. The "golden age" of Brachiopoda was the Paleozoic; they were most abundant from the Ordovician up to the Carboniferous. Many species also

FIGURE 24. Class Brachiopoda. a–d Subclass Inarticulata. e–p Subclass Articulata. a, b Order Lingulida. a_1–a_2 *Lingula*, living animal (a), Side view, 0.5x (a_1), top view (a_2), Ordovician–Recent; b, b_1 *Obolus*, 1x, Cambrian–Ordovician. c, d Order Acrotretida. c, c_1 *Discinisca*, 1x, Jurassic–Recent; d–d_4 *Crania*, 2x, pedicle valve (d_1, d_2) and brachial valve (d_3, d_4). e–e_3 Order Orthida, *Orthis*, 0.8x, Lower Ordovician. f–h Order Strophomenida. f–f_2 *Strophomena*, 0.4x, Ordovician; g–g_2 *Productus*, 0.3x, Carboniferous; h *Richthofenia* entire specimen (h) and longitudinal section (h_1), 0.3x, Permian. i–i_3 Order Pentamerida, *Gypidula*, I and II show the position of cross sections (i_2 and i_3), 0.4x, Silurian–Devonian. k, k_1 Order Rhynchonellida, *Lacunosella*, 0.4x, Upper Jurassic. l, m Order Terebratulida. l–l_2 *Terebratula*, 0.4x, Tertiary; m–m_2 *Pygope*, 0.5x, Upper Jurassic–Lower Cretaceous. n–p Order Spiriferida. n, n_1 *Spirifer*, 0.5x, Carboniferous; o, o_1 *Atrypa*, rear view (o) and cross section with sectioned spiral cones of the brachidium, 0.5x, Silurian–Devonian; p–p_2 *Spiriferina* ventral and side view, and cross section with sectioned brachidium, 0.5x, Triassic–Jurassic. Abbreviations as in figure 22 (after *Moore*).

TABLE 6.   Overview of Orders of the Articulata

Order Geological Range	Shell Structure	Shell Shape	Posterior Commis- sure	Brachio- phores, Brachidium	Various Characteris- tics and Notes	Figures
Orthida Cambrian– Permian	Mostly impunctate	Bi- or plano- convex	Straight hinge line	Brachio- phores	Open delthyrium	23, a, b 24, e
Strophomenida Ordovician– Lower Jurassic	Pseudo- punctate	Fre- quently concavo- convex	Straight hinge line	Not developed	Many aberrant forms	24, f–h
Pentamerida Middle Cambrian–Upper Devonian	Impunctate	Biconvex	Straight, short or curved	"Crura"	Spondy- lium	24, i
Rhynchonellida Middle Ordovician– Recent	Impunctate	Biconvex	Curved, cardinal margin	Crura	Shell fre- quently tri- lobate with strong ribs	23, c 24, k
Terebratulida Lower Devonian– Recent	Punctate	Biconvex	Curved, cardinal margin	Crura and calcareous loop	The shell is usually smooth, except for the growth lines	22 23, d–f 24, l–m
Spiriferida Middle Ordo- vician–Lower Jurassic	Impunctate	Biconvex	Straight, short to very long hinge line	Crura and calcareous conical spirals	Shell fre- quently with a dis- tinct sinus and swell- ing, often ribbed	23, g–i 24, n–p

occurred during the Jurassic; currently they are represented by relatively few taxa.

Phylum Mollusca (figures 25–39)

The molluscan body is, as a rule, unsegmented, consisting of a head, a body with a visceral sac, a mantle, and a ventrally located muscular foot. The mantle is a dorsal fold of the skin; it envelops the body and the pallial cavity. In most water-dwelling mollusks, the mantle (pallial) cavity contains one or more gills. Jaws may be developed in the mouth, and a radula may be developed at the beginning of the esophagus. The dorsal body cover and the mantle of most mollusks secrete a layered calcareous shell. As this shell is easily preserved, Mollusca are the most abundant of the macrofossils. The phylum Mollusca is the second largest of the invertebrate phyla. The oldest molluscan remains are from the lowest Cambrian.

The phylum Mollusca is divided into several classes, based on the structure of the shell and the presence or absence of radula.

1. Subphylum Amphineura

Upper Cambrian–Recent (figure 25, a, b)

Amphineura are bilaterally symmetrical mollusks whose body protection is either a spiny cuticle alone (class Aplacophora) or the cuticle with additional calcareous plates (class Polyplacophora). Most Amphineura live in the littoral zone.

1. Class Aplacophora (Solenogastres)

The wormlike Aplacophora (from 0.5–30 cm long but generally 1–3 cm), with about 150 recent marine species, are not known as fossils. In spite of that, it is assumed that they represent very ancient, original mollusks. The cuticle, which is protected by calcareous spicules, is considered to be a primary one and as such is an important subject of research in the development of the molluscan shell.

2. Class Polyplacophora (chitons)

Upper Cambrian–Recent (figure 25, a, b)

Polyplacophora are Amphineura whose bodies are dorsally protected by a shell of eight (rarely only seven) somewhat overlapping plates and laterally by a belt of calcareous spines or scales. The shell consists of four layers, which is a deviation from other mollusks.

The eight shell-plates are connected to a muscular foot by eight pairs of dorsoventral muscles; the mantle cavity contains numerous gills. Fossil finds of Polyplacophora are rare; usually only isolated shell plates are found.

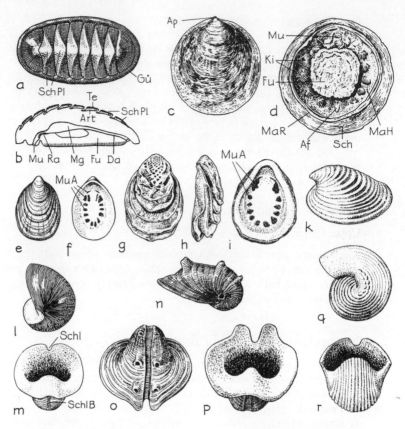

FIGURE 25. a, b Class Polyplacophora. **c–r** Class Monoplacophora. **a, b** *Chiton,* 0.7x, dorsal view **(a)** and schematic longitudinal section **(b)**, Recent. **c, d** *Neopilina,* **c** Dorsal view, 1.5x, **d** Viewed from below, the soft body shown, Recent; **e, f** *Pilina,* dorsal **(e)** and ventral **(f)** views, 0.35x, Middle Silurian; **g–i** *Tryblidium,* dorsal **(g),** lateral **(h),** and ventral **(i)** views, 0.7x, Middle Silurian; **k** *Helcionella,* 1.6x, Lower Cambrian; **l, m** *Bellerophon,* 1x, Silurian–Lower Triassic; **n–p** *Knightites,* 0.8x, Carboniferous; **q, r** *Euphemites,* 1.4x, Carboniferous. Af, anus, Ap, apex, Art, articulamentum, Da, intestine, Fu, foot, Gu, girdle, Ki, gill, MaH, mantle cavity, MaR, mantle edge, Mg, stomach, Mu, mouth, MuA, muscle scars, Ra, radula, Sch, shell, Schl, slit, SchlB, slit band (selenizone), SchlPl, shell plate, Te, tegmentum (after *Hennig, Lemche, Moore, Smith*).

2. Subphylum Conchifera

This subphylum is comprised of mollusks with a unified shell. The shell can be secondarily bivalved, as in the case of mussels.

1. Class Monoplacophora

Lower Cambrian–Recent (figure 25, c–r)

This class was established only recently, after detailed examination of specimens of *Neopilina galatheae* obtained by drag nets from 3,570-meter depths off the west coast of Costa Rica in 1952. The long-known (Paleozoic) finds of similar caplike shells had formerly been classified as gastropods.

Neopilina develops several pairs of each of various organs (six pairs of nephridia, five pairs of gills, eight pairs of dorsoventral muscles, ten pairs of lateropedal commissures, two pairs of gonads, and two pairs of auricles). This was considered an indication of original segmentation. Monoplacophora were regarded as a connecting link between the hypothetical, segmented molluscan ancestors and the true, unsegmented mollusks. However, according to newer information, the segmentation of the Monoplacophora is not conclusively primitive.

The multilayered calcareous shell of *Neopilina* is cap shaped, with eight pairs of attachment scars of dorsoventral muscles.

Tryblidium, Pilina, and other paleozoic forms also show up to eight pairs of muscle scars, arranged in a horseshoe-shaped arch. Because of this, they undoubtedly belong among the Monoplacophora. Planispiral shells of Bellerophontina, which were widely distributed during the Paleozoic (Lower Cambrian to Lower Triassic, figure 25, k–r), also have been recently interpreted as shells of Monoplacophora. They, too, have several paired muscle scars. In addition, the shells have a median slit in the lip and a slitband (selenizone).

2. Class Gastropoda (snails)

Upper Cambrian–Recent (figures 26, 27)

Gastropoda are by far the most species-rich molluscan class (approximately 105,000 recent and 15,000 fossil species). Their representatives are found in saltwater, freshwater, and terrestrial habitats.

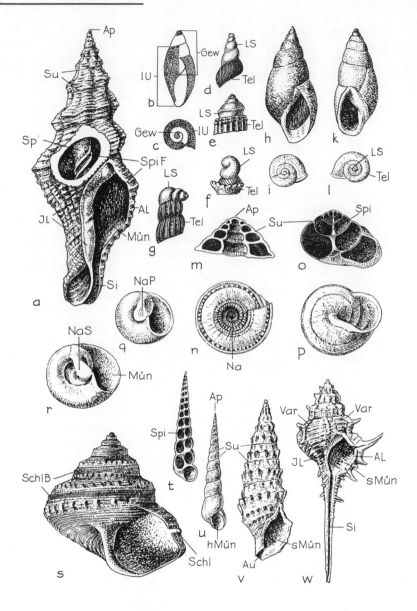

Gastropods usually have an asymmetric body protected by a spirally coiled calcareous shell. The head carries one or two pairs of sensory tentacles; in the mouth are mandibles (jaws) and a radula. The foot usually forms a wide sole for crawling.

THE SHELL (figure 26)

Usually trochospiral, more rarely planispiral, the shell consists of embryonic, larval shell (protoconch) and adult, postlarval shell (teleoconch). The protoconch is formed of one to four whorls; it is frequently sculptured differently from the teleoconch, often even differently coiled. The subsequent whorls may overlap to a greater or lesser degree (involute and evolute shells). The shell, usually dextrally coiled, begins at the apex and ends at the aperture. It is composed of the spire and the body whorl. The external line, where subsequent whorls touch each other, is called the suture. The whorls are coiled around the columella, which can be solid or hollow. The umbilicus, a funnel-shaped hollow of the columella, can be partly or completely closed by a callus. The peristome (margin of the aperture) is divided into an inner lip, adjacent to the columella, and an outer lip. Peristome may be continuous (holostomatous) or siphonostomatous. Siphonostomatous forms have either a short siphonal notch or a longer siphonal canal. Some holostomatous forms have developed a slit in the outer lip. The growth of the shell in this place is somewhat retarded. A lengthwise band running from the slit, and often sculptured differently from the rest of the shell, is designated as the slitband or selenizone.

FIGURE 26. Morphology of gastropod shells. **a** *Latirus*, last whorl partially removed to show the columella. **b, c** Dextrally coiled shell in side and plane view. **d–g** Various embryonic shells of the Gastropoda: **d, e** Orthostrophic, conical, **f** Alloiostrophic, **g** Heterostrophic. **h–l** *Phasianella* in side and plane view, dextrally coiled (**h, i**), sinistrally coiled (**k, l**). **m, n** *Architectonia*, longitudinal section (**m**), and base (**n**). **o, p** *Eobania*, longitudinal section (**o**), and base (**p**). **q, r** *Natica*, base. **s** *Pleurotomaria*, 0.5x, Jurassic. **t, u** *Turritella*, longitudinal section (**t**) and side view (**u**), 0.5x, Recent. **v** *Cerithium*, 0.5x, Recent. **w** *Murex*, 0.5x, Recent. AL, outer lip, Ap, apex, Au, siphonal notch, Gew, whorl, hMün, holostomatous aperture, IL, inner lip, LS, protoconch (embryonic shell), IU, body whorl, Mün, aperture, Na, umbilicus, NaP, callus, NaS, partial callus, Schl, slit, SchlB, slit band (selenizone), Si, siphonal canal, sMün, siphonostomatous aperture, Spi, columella, SpiF, columellar fold, Su, suture, Tel, teleoconch, Var, varix (after *Moore*).

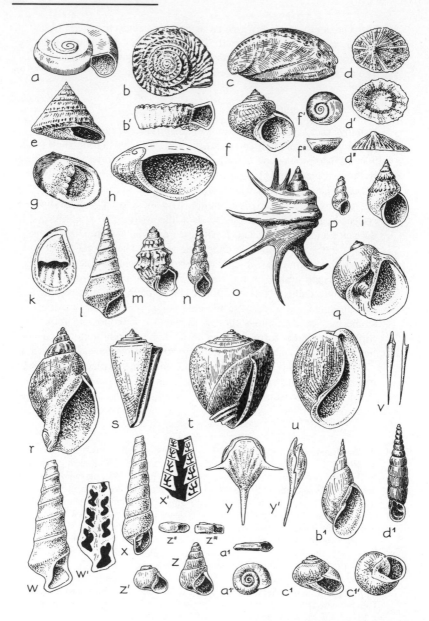

The sculpture of the gastropod shell is characterized by radially and/or spirally arranged costae, spines, beads, or bosses. Growth lines, parallel to the peristome, are easily observed in smooth-shelled forms. Many gastropods develop thickened varices (sing. varix) in the earlier peristome locations. Frequently there are secondary shell layers, as calluses or as a secondary shell covering, called the inductura. There are also many examples of resorption of the shell material. Throughout their lifetime many aquatic gastropods retain a lid (operculum) that closes the aperture. The lid is located on the posterior part of the foot and is either calcareous or composed of an organic substance. For longer rest periods (e.g., winter or drought), many terrestrial snails temporarily cover the aperture with an epiphragm.

CLASSIFICATION

The zoological system of gastropods is primarily based on the anatomical characteristics and topography of the soft body parts, mainly the gills and the nervous system, as well as the formation of the radula. In contrast, fossil gastropods can only be classified according

FIGURE 27. Class Gastropoda. a–s Subclass Prosobranchia. t–y Subclass Opisthobranchia. z–d¹ Subclass Pulmonata. a–i Order Archaeogastropoda: a *Euomphalus*, 0.5x, Silurian–Middle Permian; b *Discohelix*, 0.7x, Triassic–Cretaceous; c *Haliotis*, 0.5x, Tertiary–Recent; d *Patella*, front to the left, 0.4x, Tertiary–Recent; e *Trochus*, 0.5x, Tertiary–Recent; f *Turbo*, shell, operculum from inside (f′) and in lateral view (f″), 0.4x, Upper Cretaceous–Recent; g *Nerita*, 0.5x, Tertiary–Recent; h *Naticopsis*, 0.3x, Middle Devonian–Triassic; i *Amberleya*, 0.3x, Triassic–Jurassic. k–q Order Mesogastropoda: k *Crepidula*, 0.4x, Upper Cretaceous–Recent; l *Undularia*, 0.3x, Triassic; m *Viviparus*, 0.5x, Jurassic–Recent; n *Loxonema*, 0.4x, Middle Ordovician–Lower Carboniferous; o *Harpagodes*, 0.25x, Middle Jurassic–Upper Cretaceous; p *Hydrobia*, 1x, Upper Jurassic–Recent; q *Natica*, 0.5x, Cretaceous–Recent. r, s Order Neogastropoda: r *Buccinum*, 0.5x, Tertiary–Recent; s *Conus*, 0.5x, Tertiary–Recent. t Order Pleurocoelia, *Trochacteon*, 0.4x, Cretaceous. u Order Pleurocoelia, *Bulla*, 0.5x, Lower Jurassic–Recent. v Order Pteropoda, *Styliola*, 4x, Middle Tertiary–Recent. w *Nerinea*, shell from the outside (w) and in longitudinal section (w′), 0.4x, Lower Jurassic–Upper Cretaceous. x *Ptygmatis*, shell (x) and a longitudinal section (x′), 0.5x, Middle Jurassic–Upper Cretaceous. y *Diacria*, 2.5x, Midtertiary–Recent. z–b¹ Order Basommatophora: z–z‴ *Gyraulus*, 1.5x, Miocene; a¹ *Planorbis*, 0.7x, Tertiary–Recent; b¹ *Lymnaea*, 0.5x, Tertiary–Recent. c¹, d¹ Order Stylommatophora: c¹ *Cepaea*, 0.5x, Tertiary–Recent; d¹ *Clausilia*, 2.5x, Pliocene–Recent (after *Rieber, Zilch*).

to the morphology and structure of the shell. However, since most of the widely distributed fossil groups also have recent representatives, the system established by zoologists agrees fairly well with that used by paleontologists.

Gastropoda are divided into three subclasses: Prosobranchia, Opisthobranchia, and Pulmonata.

1.   Subclass Prosobranchia

Upper Cambrian–Recent

Gastropods with one or two gills, anterior to the heart, and with crossed pleurovisceral nerve commissures (chiastoneuric). As a rule the shell is well developed and spirally coiled. An operculum is often developed.

Most Prosobranchia live in the sea; only a few are on land or in fresh water.

1. Order Archaeogastropoda

Upper Cambrian–Recent (figures 26, s, 27, a–i)

Prosobranchia possessing one or two gills equipped along both edges with gill filaments. The inner layer of the shell is either nacreous or porcelaneous. In many cases the holostomatous aperture is interrupted by a slit or a homologous structure.

2. Order Mesogastropoda

Ordovician–Recent (figures 26, m–n, q–r, t–v, 27, k–q)

Prosobranchia with one gill that is equipped with filaments on one side only. The inner layer of the usually well-developed shell is always porcelaneous. The aperture typically shows a short siphonal notch. However, numerous mesogastropods are holostomatous.

3. Order Neogastropoda

Ordovician–Recent (figures 26, a, w, 27, r–s)

Both the anatomy of the soft parts and the shell structure are the same as in the Mesogastropoda. Typical shells are siphonostomatous, with a longer siphonal canal. Most neogastropods are marine.

2. Subclass Opisthobranchia

Carboniferous–Recent (figure 27, t–y)

Gastropods with one gill and one atrium located laterally behind the heart. The pleurovisceral commissures are not crossed (euthyneuric). Disregarding a few exceptions (Acteonidae and Nerinacea), the general trend is toward reduction of the shell. Most modern Opisthobranchia inhabit the sea and are naked snails.

The more common fossil forms belong to Acteonidae and Nerinacea. *Acteonella* and *Trochacteon* (figure 27, t), the most abundant representatives of Acteonidae (order **Pleurocoelia**) from the Upper Cretaceous, have a thick shell with three characteristic columellar folds. Most of the Nerinacea (figure 27, w, x), limited to the Jurassic and Cretaceous, have sharply conical, thick shells. The amount of room inside the shell is generally limited by the presence of spiral folds. Many representatives of Nerinacea and Acteonidae were adapted to life in shallow waters of lagoons and frequently occurred there in large masses.

Representatives of the order **Pteropoda** (figure 27, v, y), documented since the Upper Cretaceous, have a foot that is transformed into a fin-shaped swimming organ. The shell, much reduced, is pointed, conical, or cap-shaped. Pteropods are planktonts of the high seas. Their empty shells are constituents of the pteropode-ooze of the 1,000–2,700 meter–deep oceans.

The shell of *Berthelina* (Eocene–Recent, order **Saccoglossa**) consists of two valves held together by an adductor muscle.

3. Subclass Pulmonata

Uppermost Jurassic–Recent (figure 27, z–d^1)

Gastropods in which the ceiling of the mantle cavity is transformed into an air sac or lung. Gills are absent, and the pleurovisceral commissures do not cross (euthyneuric). Most Pulmonata inhabit dry land, some fresh water, and a few, the littoral zone.

All members of the order **Basommatophora** (figure 27, z–b^1) have a well-developed shell; their eyes are located on the bases of the tentacles.

Best known is the genus *Gyraulus,* because it occurred in great abundance in the Steinheim basins (South Germany). Several distinct shapes of the shell are known. They are interpreted in part as varieties of one species, in part as separate species.

Members of the order **Stylommatophora** (terrestrial lung snails) (figure 27, c¹–d¹) have eyes located on the distal ends of retractable tentacles. Their shells are mostly thin-walled and holostomatous. In numerous forms the shell is partly or completely reduced. Most Stylommatophora live on dry land. Because many of them are adapted to a distinct, narrow microclimate, they may be used as indicators of climatic conditions. For example, the presence of certain snails in the loess indicates that the climate was cold when the loess was formed.

3. Class Scaphopoda

Ordovician–Recent (figure 28)

Marine Conchifera with a dorsoventrally stretched body. The mantle grows together ventrally and secretes a conical, tubular shell. The head is ringed by two bunches of food-gathering tentacles (captacula). Scaphopods stick slantwise in soft sediments from which they collect foraminifers by the captacula. The most important genus is *Dentalium* (Triassic–Recent).

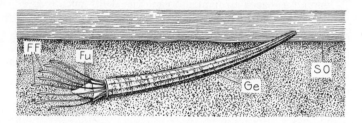

FIGURE 28. Scaphopoda. *Dentalium* in its natural habitat, 1x. FF, captacula, Fu, foot, Ge, shell, SO, top of the sediment layer (after *Hennig*).

4. Class Bivalvia (mussels, Lamellibranchia)

Lower Cambrian–Recent (figures 29–30)

Organization: Bivalvia are normally bilaterally symmetrical. The two-part mantle surrounds a lateral gill cavity and a centrally located body. A head is not differentiated. Mouth and anus lie at opposite ends, and a muscular foot is located ventrally toward the front end. The soft body is protected by a bivalval shell, the valves of which are held together by a dorsal ligament. Most mussels are elongated, with flattened sides. All live in the water, mostly in the sea. Recent mussels are vagile or sessile, living on or in the sediment. A pseudoplanktonic way of life has been postulated for several fossil forms.

SHELL (figure 29)

The oldest part of the shell, the beak, is located dorsally on each valve and is ringed by concentrically arranged growth lines. The beaks usually point forward (are prosogyre), but often they may be directed toward each other (orthogyre) or face the rear (opisthogyre). The shell surface may be either smooth or variously sculptured, in addition to the concentric growth lines. Many mussels have a distinctly delineated area, designated as the lunule, ahead of the beak. A corresponding area that may occur to the rear of the beak on one or both valves is designated as the escutcheon. Extensions of the valve ahead or behind the beak are called auricles. Anterior to the beak may be a small slit or notch in the valve margin, the byssal opening.

The ligament, which is a noncalcified part of the shell, is firmly connected to both valves, either below or behind the beak. It may be either in one piece or composed of several parts. In addition to the flexible, elastic ligament is a nonelastic, thickened ligament in the dorsal region. When there is no pressure on the elastic ligament, it holds both valves slightly open. Contraction of the transverse adductor muscles closes the valves and bends the ligament. The adductor muscle scars are usually easily discernible, even in fossil materials. The musculature consists of an anterior and a posterior adductor; sometimes only the posterior adductor is developed. Homomyarian mussels have two nearly equally developed adductors, heteromyarian have a smaller anterior adductor, and monomyarian have only the posterior muscle. The line, at which the mantle is marginally attached to the myostracum of the shell, is called the pallial line. It joins the two adductor scars and runs parallel with the edge of the shell (integripalliate) or has a more or less deep indentation (pallial sinus) toward the rear of the shell (sinupalliate).

Hingeline and Dentition (figure 29)

When closed, the two valves touch along the line of the commissure. The dorsal commissure of many mussels is widened and carries toothlike processes and corresponding sockets. These teeth and sockets prevent sideways slipping of the closed valves. The structure of the hingeline, also called dentition, is an important taxonomic characteristic.

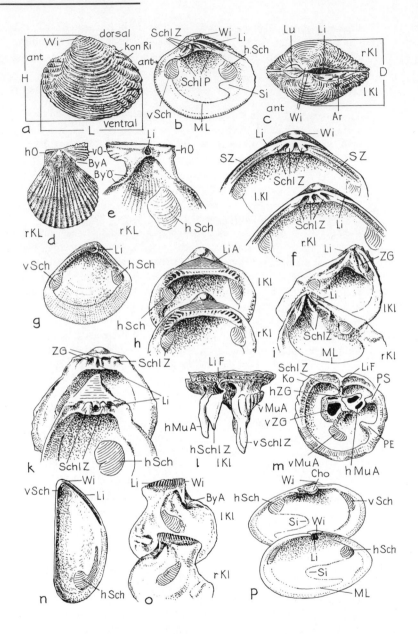

Taxodont dentition is characterized by numerous similar teeth and sockets. The position of the teeth and the presence or absence of a ligament area below the beak distinguish several types of dentition: ctenodont (teeth converging toward the center of the valve, ligament area absent), actinodont (teeth converging toward the beak), and pseudoctenodont (teeth converging toward the center of the valve, ligament area present).

Heterodont (or teleodont) dentition is characterized by up to three short cardinal teeth on a hinge plate below the beak and by corresponding sockets. In addition, there are anterior, and frequently also posterior, ridgelike lateral teeth and corresponding elongated sockets. The number of cardinals and laterals indicates further subtypes, such as the cyrenoid or the lucinoid dentition. The heterodont dentition can be described by formulas; however, establishing a formula presupposes a knowledge of the ontogenic development of the hinge.

Praeheterodont dentition is characterized by two short teeth and sockets below the beak and by one or two markedly longer teeth and sockets behind the beak.

Schizodont dentition is characterized by only two large, diverging, grooved teeth below the beak. The rear tooth of the right valve is triangular and very wide.

Mussels with pachyodont dentition may have up to two often enormously large and modified teeth and corresponding sockets. Most pachyodont mussels have aberrant shells as well.

Dysodont dentition is characterized by an absence of distinct teeth and sockets, and isodont dentition by two teeth and sockets arranged symmetrically around the ligament. Typical desmodont den-

FIGURE 29. Class Bivalvia. Shell morphology and hinge types (dentition given in parentheses). a–c *Venus* (heterodont), d, e *Chlamys* (dysodont), f *Corbicula* (heterodont), g *Nucula* (taxodont–ctenodont), h *Glycimeris* (taxondont–pseudoctenodont), i *Myophorella* (schizodont), k *Spondyllus* (isodont), l, m *Hippurites* (pachyodont), n *Mytilus* (dysodont), o *Isognomon* (dysodont), p *Mya* (desmodont). ant, anterior, Ar, escutcheon, ByA, byssal notch, ByÖ, byssal opening, Cho, chondrophore, D, thickness, H, height, hMuA, posterior adductor muscle scar, hO, posterior auricle, hSch, posterior adductor, hSchlZ, posterior hinge tooth, hZG, posterior socket, Ko, commissure, konRi, concentric costae, L, length, Li, ligament, LiA, ligament area, LiF, ligament ridge, lKl, left valve, Lu, lunule, ML, pallial line, PE, processus E, PS, pillar S, rKl, right valve, SchlP, hinge plate, SchlZ, hinge tooth, Si, pallial sinus, SZ, lateral tooth, vMuA, anterior adductor muscle scar, vO, anterior auricle, vSch, anterior adductor, vSchlZ, anterior hinge tooth, vZG, anterior socket, Wi, beak, ZG, socket.

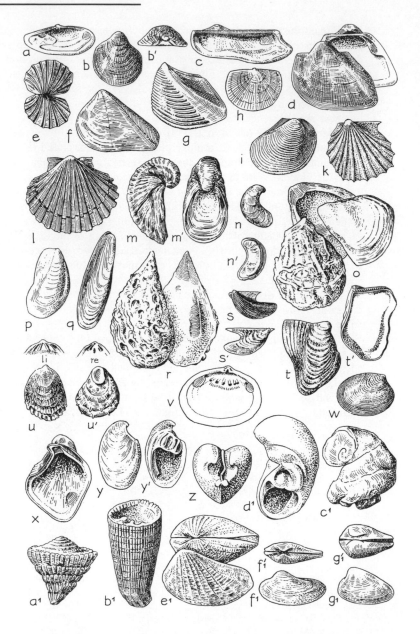

tition is also characterized by a lack of distinct teeth. A chondrophore, a spoon-shaped structure below the beak, is frequently developed. It serves for attachment of the inner ligament (resilium).

Shell Composition (table 7)

The laminar shell of mussels is secreted by the mantle, as in other Conchifera. The space between the shell and the mantle epithelium (the extrapallial cavity) is filled with a liquid. The soft body touches the shell only at muscle attachments (e.g., adductors), and the pallial line. The composition of the extrapallial-cavity fluid is one of the factors affecting the shell composition and structure.

Periostracum, the always thin outermost layer of the shell, is secreted by the most external of the three mantle folds. It is composed of tannin-containing protein, conchiolin, and serves as a substrate for the subsequent growth of the shell as well as for its protection from secondary dissolution. Underneath the periostracum lies a much thicker calcareous shell that can vary in its structure but usually comprises layers of aragonite and/or calcite alternating with thin conchiolin lamellae.

FIGURE 30. Class Bivalvia. Subclass Palaeotaxodonta (**a**); Subclass Cryptodonta (**b**, **e**); Subclass Pteriomorpha, order Arcoida (**c**, **d**); Order Mytiloida (**p–r**); Order Pterioidea (**h–o**, **s–u**); Subclass Palaeoheterodonta (**f**, **g**); Subclass Heterodonta, order Veneroida (**v–x**); Order Hippuritoidea (**y–d¹**); Subclass Anomalodesmata (**e¹–g¹**). **a** *Yoldia*, 1x, Cretaceous–Recent; **b** *Cardiola*, 0.5x, Silurian–Devonian; **c** *Arca* 0.4x, Upper Cretaceous–Recent; **d** *Cucullaea*, 0.4x, Jurassic–Recent; **e** *Buchiola*, 5x, Silurian–Devonian; **f** *Myophoria*, 0.6x, Triassic; **g** *Trigonia*, 0.3x, Middle Triassic–Upper Cretaceous; **h** *Daonella*, 0.3x, Middle–Upper Triassic; **i** *Posidonia*, 0.5x, Lower Carboniferous–Upper Jurassic; **k** *Oxytoma*, 0.7x, Upper Triassic–Cretaceous; **l** *Pecten*, 0.2x, Tertiary–Recent; **m** *Gryphaea*, 0.3x, Jurassic–Cretaceous; **n** *Exogyra*, 0.6x, Middle Jurassic–Upper Cretaceous; **o** *Ostrea*, 0.3x, Cretaceous–Recent; **p** *Modiolus*, 0.5x, Devonian–Recent; **q** *Lithophaga*, 0.5x, Middle Tertiary–Recent; **r** *Trichites*, 0.2x, Middle Jurassic–Lower Cretaceous; **s** *Rhaetavicula*, 0.8x, Upper Triassic; **t** *Inoceramus*, 0.2x, Lower Jurassic–Upper Cretaceous; **u** *Plicatula*, 0.7x, Middle Triassic–Recent; **v** *Babinka*, 1.1x, Middle Ordovician; **w** *Astarte*, 0.5x, Jurassic–Recent; **x** *Congeria*, 0.2x, Tertiary; **y** *Megalodon*, 0.25x, Devonian–Triassic; **z** *Neomegalodon*, 0.2x, Upper Triassic; **a¹** *Radiolites*, 0.25x, Upper Cretaceous; **b¹** *Hippurites*, 0.25x, Upper Cretaceous; **c¹** *Requienia*, 0.2x, Cretaceous; **d¹** *Diceras*, 0.3x, Upper Jurassic; **e¹** *Pholadomya*, 0.35x, Upper Triassic–Recent; **f¹** *Pleuromya*, 0.4x, Triassic–Lower Cretaceous; **g¹** *Gresslya*, 0.4x, Jurassic. re, right valve, li, left valve (after *Rieber, Moore,* and *Teichert*).

TABLE 7.   Frequently Occurring Structural Types of Calcareous
Shells of Bivalves

Structure	Compound	Microstructure	Occurrence in the Calcareous Shell
Nacreous	Aragonite	Platelets parallel to the inner surface, separated by conchiolin layers	Only as inner shell layer
Lamellar	Calcite	Calcite flakes 10–15 micrometers long and 0.1–0.6 micrometers thick	Entire shell
Prismatic	Aragonite	Aragonite prisms 0.05–0.1 mm in diameter	Outer shell layer
Prismatic	Calcite	Calcite prisms approx. 0.1 mm in diameter	Outer shell layer
Myostracum	Aragonite	Prisms	Muscle attachment scars
Cross-lamellar structure	Aragonite	Crossing aragonite lamellae, consisting of extremely fine (approx. 8 micrometers long and 1 micrometer thick) layers	Entire shell

The calcareous layer of mussels with calcite shells is frequently
fossilized; in contrast, the unstable aragonite of mussels that were
once housed in aragonitic shells is often dissolved during the
diagenesis (e.g., oysters or pectinids).

CLASSIFICATION (figure 30, table 8)

The generally employed classification of mussels is based on the
structure of the shell, hinge, and adductor muscles as well as on the
ligaments, pallial line, and gills.

To date, the stratigraphically oldest known mussel, tiny (1.2
mm) *Pojetaia runnegari* from Australian Lower Cambrian, is classified with the Palaeotaxodonta. In contrast, *Fordilla,* also Lower

TABLE 8.   Overview of the Subclasses and of Some Orders of Mussels

Subclass Order Geological Range	Hinge Dentition	Shell Structure (if known)	Adductor Muscles, Pallial Line	Figure(s)
Palaeotaxodonta Lower Cambrian– Recent	Taxodont- ctenodont	Aragonitic prism layer and nacreous structure	Homomyarian, mostly integripalliate	29, g 30, a
Cryptodonta Ordovician– Recent, predominantly Paleozoic	Without distinct dentition	Aragonitic	Homomyarian, integripalliate	30, b, e
Pteriomorphia Lower Cambrian– Recent				
Arcoida Ordovician– Recent	Taxodont- pseudoctenodont	Cross-lamellar structure	Homomyarian, integripalliate	29, h 30, c, d
Mytiloida[1] Cambrian– Recent	Dysodont	Thick calcareous prismatic layer and thin nacreous structure	Hetero- or monomyarian, entire margined	29, n 30, p–r
Pterioida Ordovician– Recent	Dysodont or isodont	Either a thick calcareous prismatic layer with a thin nacreous layer, or a calcitic lamellar structure	Hetero- or monomyarian, entire margined	29, d, e, k, o 30, h–o, s–u
Palaeoheterodonta Ordovician– Recent	Actino-, prehetero-, or schizodont	Aragonitic prismatic layer and nacreous structure	Homomyarian, integripalliate	29, i 30, f, g
Heterodonta Ordovician– Recent	Hetero-, pachyo-, or desmodont	Frequently cross-lamellar; in Pachyodonta with a thick calcitic outer layer	Homomyarian, more rarely heteromyarian, integri- or sinupalliate	29, a–c, f, l, m, p 30, v–d[1]
Anomalodesmata Ordovician– Recent	Desmodont	Cross-lamellar ± nacreous	Homomyarian, sinupalliate	30, e[1]–g[1]

[1]Mytiloida, together with the Modiomorphida (here designated as an order of the Palaeoheterodonta) were recently placed in a separate class, Isofilibranchia.

Cambrian, is regarded as a member of the order Mytiloidea. Both have numerous common characteristics and are quite acceptable as ancestors of most of the higher bivalvian taxa.

Most fossil heterodont mussels are of minor importance; however, among the pachyodonts there are numerous important ones. The shapes of most Cretaceous pachyodont mussels are aberrant to such a degree that originally they were not even recognized as mussels. *Radiolites* and *Hippurites* (figure 30, a[1], b[1]) and many related forms occurred in great abundance in warm, shallow seas of the Upper Cretaceous and frequently created extensive reef banks. The Upper Cretaceous pachyodont mussels are often grouped as rudistids. The pachyodont mussels of the Triassic (*Neomegalodon* and related forms) and the Upper Jurassic (*Diceras*) were abundant and widely distributed in the lagoon regions of carbonate platforms.

Desmodont representatives of the Heterodonta are thin-shelled mussels that live either, like the *Mya,* buried in soft sediments or, like the boring mussel, *Pholas,* in hard substrates. Their shells are posteriorly (in part also anteriorly) dehiscent (widely open).

ECOLOGY

Mussels are well suited for paleoecological investigations, because their rich variety reflects successful adaptation to a wide range of living conditions. Since the position of the soft parts can be partially inferred from the locations of attachment scars and openings of the hard shell, it is possible to study even the fossil mussels functionally. Most mussels live in the continental shelf region, which makes them accessible to direct observation. In addition, numerous families described on the basis of recent mussels were already represented by many very similar forms during the Mesosoic. This makes it possible to arrive at reliable conclusions regarding the ecology of many fossil mussels.

5. Class Cephalopoda

Upper Cambrian–Recent (figures 31–38)

The exclusively marine Cephalopoda represent the highest level of molluscan development. Their bilaterally symmetrical bodies consist of a head with tentacles, an intestinal sac, and a mantle. The calcareous shell of the original cephalopods is located externally; the shell of derived forms is overgrown by the soft body, forming an inner skeleton, or is reduced. The ventrally located mantle cavity opens to the outside by means of a funnel. Recent cephalopods have strong jaws and a radula.

Modern Cephalopoda are active, carnivorous, mostly predators, and more rarely scavengers. Some are quite agile swimmers.

Fossil cephalopods, especially the diverse group of Ammonoidea, serve as excellent index fossils for the period from the Devonian to the end of the Cretaceous.

Recent Cephalopoda are classified according to the number of gills as **Tetrabranchiata** or **Dibranchiata,** or according to the shell development as **Ectocochlia** (external shell) or **Endocochlia** (internal or reduced shell). Since *Nautilus,* the only recent representative of the Tetrabranchiata, is also the only living representative of the Ectocochlia, it is generally assumed that the fossil ectocochliate and endocochliate cephalopods were also tetra- and dibranchiate, respectively.

MORPHOLOGY OF THE SHELL

The ectocochliate shell is basically a narrow, conical tube. Its rear part, the phragmocone, is partitioned into chambers by transverse walls, the septa. The soft body is contained in the living chamber, which is the undivided front part. The edge of the opening is termed the apertural margin. The siphon, a thin cord of the soft body, runs the length of the shell, passing through each septum via a siphonal foramen. Edges of the siphonal foramina are turned either anteriorly or posteriorly, forming septal necks. The siphon can be located either centrally or along the side. The outer shell is multilaminar: underneath a thin periostracum is an aragonite prismatic layer, and further inside is a nacreous (mother of pearl) layer, also prismatic. The structure of the septa is prismatic as well.

The narrow conical shell may be straight, bent, or spirally involuted.

The chambers of the phragmocone of a living animal are filled with gas, which provides great buoyancy.

The shell of most endocochliate cephalopods is enveloped in a fold of the body, acting virtually as an inner skeleton. Many Endocochlia have a shell that is greatly reduced; it may even be absent entirely.

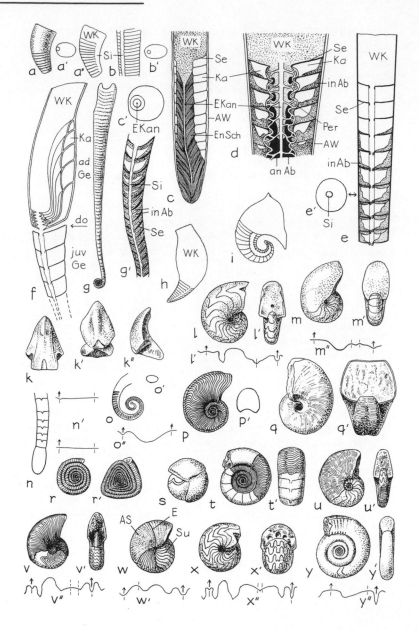

Ectocochliate Cephalopoda (Tetrabranchiata)

Tetrabranchiata are divided into several subclasses and orders according to the morphological characteristics of their shells (cf. table 9).

Subclass Ammonoidea

MORPHOLOGY OF THE SHELL (figure 32)

The tubular shell of most ammonoids is planispirally coiled, although in some representatives it can form a loosely or tightly coiled planispiral, or even a helicoid spire. A few are nearly straight, others are irregularly bent or coiled. The conch (shell) of the Ammonoidea grows from a tiny protoconch (0.5 mm in diameter); the attached chambered phragmocone of an adult may be comprised of several whorls. Length of the outermost part—the living chamber—depends

FIGURE 31. Important ectocochliate Cephalopoda. **a, b, e–m, q** Subclass Nautiloidea. **c** Subclass Endoceratoidea. **d** Subclass Actinoceratoidea. **n** Subclass Bactritoidea. **o, p, r–y** Subclass Ammonoidea. **a** *Plectronoceras*, 1.5x, Upper Cambrian; **b** *Ellesmeroceras*, 0.5x, Lower Ordovician; **c** *"Endoceras,"* schematic longitudinal section, 0.4x, Ordovician; **d** *"Actinoceras,"* schematic longitudinal section, 0.5x, Ordovician; **e** *"Orthoceras,"* schematic longitudinal section, 0.5x, Ordovician–Silurian; **f** *Ascoceras*, 0.6x, Silurian; **g** *Lituites*, 0.15x, Ordovician; **h** *Oncoceras*, 0.3x, Ordovician; **i** *Phragmoceras*, 0.2x, Silurian; **k** *Palaeoteuthis*, calcified part of the upper mandible of *Nautilida* of the Middle Jurassic, 1x; **l** *Aturia*, showing the septum attachment line, 0.3x, Early Tertiary; **m** *Nautilus* showing the septum attachment line, 0.1x, Recent; **n** *Bactrites*, with suture, 8x, Devonian; **o** *Anetoceras*, 0.5x, Lower Devonian; **p** *Mimagoniatities*, 0.8x, Lower to Middle Devonian; **q** *Germanonautilus*, 0.1x, Triassic; **r** *Soliclymenia*, 0.25x, Upper Devonian; **s** *Epiwocklumeria*, 0.3x, Upper Devonian; **t** *Anarcestes*, 0.3x, Lower to Middle Devonian; **u** *Agoniatites*, 0.1x, Middle Devonian; **v** *Manticoceras*, with suture line, 0.4x, Middle Devonian; **w** *Cheiloceras*, with suture line, 0.5x, Upper Devonian; **x** *Goniatites*, with suture line, 0.7x, Lower Carboniferous; **y** *Clymenia*, with suture line, 0.3x, Upper Devonian. adGe, adult shell, anAb, annulosiphonate deposits, AS, growth lines, AW, outer wall of the shell, E, constrictions, EKan, endosiphonal canal, EnSch, endosiphodal sheath, inAb, intracameral calcareous partitions, juv Ge, juvenile shell, Ka, chamber, Per, perispatium, Se, septum, Si, siphon, Su, suture, WK, living chamber. On the suture lines, the left arrow marks the dorsal; the right arrow, the ventral side; and the short dashes, the umbilical seam (after *Erben, Fischer, Rieber, Moore*).

TABLE 9. Overview of the Subclasses and Orders of Ectocochliate Cephalopoda

Subclass Order Geological Range	Type and curvature of the Tubular Shell	Suture Line; Spacing of Septa
Nautiloidea Upper Cambrian– Recent	Brevicone to longicone[1]; orthocone, curved or plan- ispirally coiled	Straight or only slightly curved; narrow to wide
Ellesmerocerida Upper Cambrian– Upper Ordovician	Brevicone or longicone; cyrtocone or orthocone	Straight; close
Orthocerida Ordovician–Triassic	Longicone; ortho- to weakly cyrtocone	Straight; wide
Ascocerida Ordovician–Silurian	Juvenile orthoceridlike, adult inflated; weakly cyrtocone	Reaching dorsally far forward in adults; close
Oncocerida Ordovician–Carbonif- erous	Brevicone with an inflated living chamber; weakly cyrtocone	Straight; close
Discosorida Middle Ordovician– Upper Devonian	Brevicone; gyrocone	Straight; close
Tarphycerida Ordovician–Silurian	Longicone; juvenile shell or the entire shell plan- ispirally coiled	Straight; moderately wide
Nautilida Devonian–Recent	Longicone; gyrocone to strongly planispirally involute	Straight, to multiple folds; relatively wide
Endoceratoida Ordovician	Longicone; orthocone	Straight; relatively wide
Actinoceratoida Ordovician– Carboniferous	Longicone; orthocone	Straight; relatively wide
Bactritoidea Upper Silurian–Upper Permian	Longicone; ortho- to gyro- cone or loosely plan- ispirally coiled	Ventral lobe, ± lateral lobe; relatively wide
Ammonoidea Lower Devonian– Upper Cretaceous	Longicone; usually tightly planispirally involute	Heavily lobed to very convolute; moderately wide

[1]Special terminology: brevicone = short conical, longicone = long conical, cyrtocone = slightly curved, gyrocone = sharply curved, eurysiphonate = with a wide siphon, stenosiphonate = with a narrow

Structure of the Siphon; Its Position	Various Characteristics; Mode of Life	Figure(s)
Stenosiphonate[1]; central or subcentral, seldom marginal	Partially with intracameral[1]; more rarely with endosiphonal[1] calcareous deposits; predominantly benthic	31, a, b, e–m, q
Stenosiphonate; marginal	Small, oldest Cephalopoda; benthic	31, a, b
Stenosiphonate; central or subcentral	Some with intracameral, more rarely with endosiphonal calcareous deposits; predominantly benthic	31, e
Stenosiphonate; marginal	The juvenile shell is broken off at a truncation locus; adults nektonic.	31, f
Stenosiphonate; ± marginal	Apertural margin often much constricted; benthic	31, h
Moderately wide; ± marginal	Some with endosiphonal calcareous deposits; benthic	31, i
Stenosiphonate; subcentral	Some with intracameral calcareous deposits; benthic	31, g
Stenosiphonate; central to subcentral	Shells from smooth to heavily sculptured; nektonic, living near the bottom	31, l, m, q
Eurysiphonate; marginal	Siphon with endosiphonal calcareous sheaths and central canal; benthic	31, c
Eurysiphonate; marginal	Complicated endosiphonal calcareous deposits with a system of canals, also intracameral deposits; benthic	31, d
Stenosiphonate; ventral	Small undifferentiated forms; nektonic and/or benthic	31, n
Stenosiphonate; ventral (Clymeniida dorsal/ internal)	Numerous, differently sculptured forms; nektonic; to various extent, bottom dependent	31, o–y, 32–36

siphon, endosiphonal and intra-cameral deposits = primary cal-careous deposits in the siphon or in the chambers

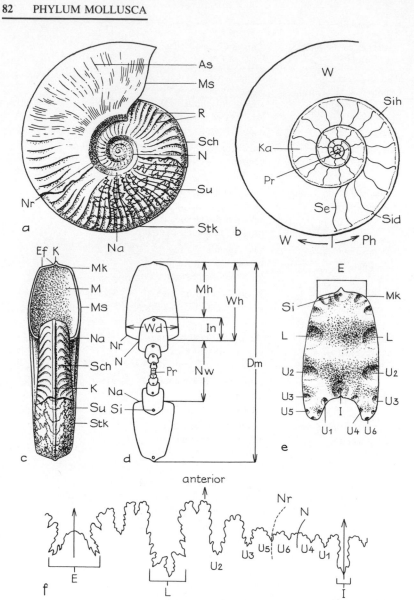

on the diameter of the coil. The shape and sculpturing of adult and juvenile conchs may be somewhat different.

The apertural margin of fully developed conchs can take many different forms. Usually it is simply rounded, but in the small, stratigraphically younger forms it is often narrowed and restricted by lateral apophyses; it may have an external rostrum.

Most shells are distinctly sculptured. The most common elements of the sculpturing are ribs, knobs, ridges, and furrows. The siphon of the Ammonoidea is always narrow and located on the dorsal (external) side of the shell, in contrast to the other Ectocochlia. Only in the Clymeniida is it located internally.

The Suture (figures 32, 33)

Edges of the septa are variously bent and lobed, giving a distinct contour to the line where the septum meets the outer shell. This line, termed the suture, is, in the stratigraphically older forms, relatively simple, but during the phylogeny it becomes more and more complicated. Both onto- and phylogenetically first-occurring folds are posteriorly convex lobes and anteriorly convex saddles. They are later frilled, forming lobules (jags directed backward) and folioles (jags directed forward) of increasing complexity.

The shape of the suture, though it can be observed only when the shell is absent, proved to be taxonomically important. The lobes are named according to their position and the order of ontogenic origin. All Ammonoidea (the only exception being the Clymeniida) show at least three elements: an *external lobe* (E), a *lateral lobe* (L) located symmetrically on each side, and an *internal lobe* (I), located on the inside. In more highly differentiated ammonoids, one or more *adventitious lobes* (A) and/or *umbilical lobes* (U) may occur.

FIGURE 32. Terminology and measurements of ammonoid shells, using the example of *Ludwigia*, Middle Jurassic. **a** Specimen with part of the shell removed, 0.5x, **b** Longitudinal section, 0.5x, **c** Frontal view, 0.5x, **d** Median cross section, 0.5x, **e** Septum, anterior plane view, 1x, **f** Suture line, right side, 1.3x. As, growth line, Dm, diameter, E, external (ventral) lobe, Ef, external (ventral) surface, I, internal (dorsal) lobe, In, involution, K, keel, Ka, chamber, L, lateral lobe, M, aperture, Mh, height of the aperture, Mk, edge of the apertural margin, Ms, apertural margin, N, umbilical seam, Na, umbilical wall, Nr, umbilical rim, Nw, umbilical width, Ph, phragmocone, Pr, proloculus, R, ribs, Sch, Shell, Se, septium, Si, siphon, Sid, septal neck, Sih, siphuncle, Stk, steinkern, Su, suture, U_1-U_6, umbilical lobes, W, body chamber, Wd, whorl width, Wh, whorl height.

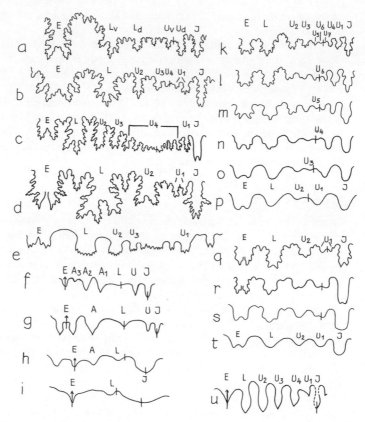

FIGURE 33. Ontogenic development of Ammonoidea sutures. Drawn to comparable sizes. **a** Order Ancyloceratida. **b** Order Ammonitida. **c** Order Phylloceratida. **d** Order Lytoceratida. **e** Order Ceratitida. **f–h** Order Goniatitida. **i** Order Anarcestida. **k–p** Order Ammonitida. **q–t** Order Lytoceratida. **u** Order Prolecanitida. **a** *Douvilleiceras*, Lower Creataceous; **b** *Amaltheus* (typical ammonitic suture), Lower Jurassic; **c** *Holcophylloceras*, Jurassic–Cretaceoous; **d** *Lytoceras*, Lower Jurassic; **e** *Ceratites* (typical ceratitic suture), Middle Triassic; **f** *Discoclymenia*, Upper Devonian; **g** *Goniatites* (typical goniatitic suture), Lower Carboniferous; **h** *Cheiloceras*, Upper Devonian; **i** *Subanarcestes*, Middle Devonian; **k–p** *Haploceras*, Upper Jurassic; **q–t** *Lytoceras*, Lower Jurassic; **u** *Prolecanites*, Lower Carboniferous. A, adventitious lobe, E, external (ventral) lobe, I, internal (dorsal) lobe, L, lateral lobe, Ld, dorsal part of the lateral lobe, Lv, ventral part of the lateral lobe, U, umbilical lobe. The representations of ontogenic development of the suture of *Haploceras* and *Lytoceras* show the first, primary suture above that of an adult shell (after *Kullmann* and *Wiedmann*, *Schindewolf*).

MANDIBLE AND RADULA (figure 34)

Single, horny plates (anaptychi) and pairs of calcitic plates (aptychi) have been known since long ago. As the contour of a pair of aptychi usually closely agrees with the shape of the aperture of the corresponding ammonite shell, they were earlier considered to be cover structures. However, thorough modern studies leave no doubt that both anaptychi and aptychi represent the lower mandible of the Ammonoidea. The upper mandible consisted only of a horny substance and therefore remained preserved extremely rarely. In addition, the presence of a radula has been proven for many ammonoids.

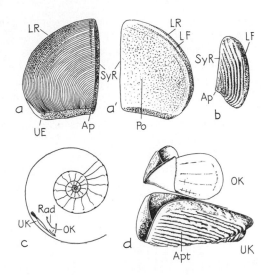

FIGURE 34. Aptychi of the Ammonoidea.
a *Laevaptychus* 0.5x, Upper Jurassic; **b** *Lamellaptychus,* 0.5x, Upper Jurassic; **c** Longitudinal section of *Eleganticeras,* 0.3x, Lower Jurassic; **d** Model of the mandibles of *Hildoceras,* Upper Lias; 1.2x. Ap, apex, Apt, aptychus, LF, lateral facet, LR, lateral margin, OK, upper mandible, Po, pores, Rad, radula, SyR, symphyseal margin, UE, umbonal corner, UK, lower mandible (after *Rieber, Lehmann*).

BIOLOGY

Ammonoids were slow-moving nektonic organisms and, as such, probably lived near the bottom. They fed on small organisms such as foraminifers, ostracodes, small ammonites, and crinoids.

Many stratigraphically younger ammonoid species were distinctly dimorphic, large forms with a simple apertural margin and small forms with apophyses definitely belonging to the same species. This is usually regarded as sexual dimorphism. Studies of commensals that lived on the ammonites indicate that ammonite shells grew over a period of several years.

CLASSIFICATION OF THE AMMONOIDEA (figures 35, 36, table 10)

The shape and morphogenesis of the suture and the position of the siphon are used to classify the ammonoid orders. Further division within the orders is based on the morphogenesis of the suture and on the form and sculpturing of the conch.

PHYLOGENY OF THE AMMONOIDEA

The main trend in the ammonoid phylogeny is the increase in complexity of the suture. Increasingly complex, multilobate sutures evolved from the trilobate suture of the oldest, planispirally coiled ammonoids. Several distinct stages of this development are marked by the goniatite, ceratite, and ammonite suture types. Progressive evolution is also shown in the development of the siphuncle from the retrosiphonate to the prososiphonate type and in the development of the curvature of septa from concave to convex. In addition, there is a phylogenic increase in size, from the oldest forms, which had tests only a few centimeters in diameter, to the 2.5-m *Pachydiscus* (Upper Cretaceous). It may be accepted as certain that Anarcestida were derived from Bactritoidea and Goniatitida from Anarcestida. Clymeniida and Prolecanitida conceivably developed from either Anarcestida or Goniatitida. Ceratitida, and probably Phylloceratida as well, arose from Prolecanitida. The Jurassic and Cretaceous

FIGURE 35. Important representatives of the Ammonoidea. **a** Order Prolecanitida. **b–h** Order Ceratitida. **i, k** Order Phylloceratida. **l–v** Order Ammonitida. **a** *Prolecanites*, 0.5x, Lower Carboniferous; **b** *Xenodiscus*, 0.4x, Upper Permian; **c** *Ceratites*, 0.2x, Middle Triassic; **d** *Trachyceras*, 0.5x, Middle Triassic; **e** *Tropites*, 0.5x, Upper Triassic; **f** *Arcestes*, f₁, preserved shell, f₂, steinkern, 0.15x, Middle to Upper Triassic; **g** *Choristoceras*, 0.5x, Upper Triassic; **h** *Cladiscites*, 0.3x, Upper Triassic; **i** *Monophyllites*, 0.25x, Middle to Upper Triassic; **k** *Phylloceras*, 0.2x, Jurassic–Cretaceous; **l** *Psiloceras*, 0.4x, Lower Lias; **m** *Arietites*, 0.05x, Lower Lias; **n** *Hildoceras*, 0.4x, Upper Lias; **o** *Harpoceras*, 0.2x, Upper Lias; **p** *Dactylioceras*, 0.4x, Upper Lias; **q** *Otoites*, 0.4x, Middle Jurassic; **r** *Spiroceras*, 0.5x, Middle Jurassic; **s** *Perisphinctes*, 0.2x, Upper Jurassic; **t** *Oppelia*, 0.3x, Middle Jurassic; **u** *Hoplites*, 0.3x, Lower Cretaceous; **v** *Acanthoceras*, 0.25x, Upper Cretaceous (after *Moore, Rieber*).

TABLE 10. Overview of the Orders of Ammonoidea

Order Geological Range	Suture Type of Adult Shell; First Suture	Notable Characteristics of the Suture	Shape of the Shell, Other Characteristics	Figure(s)
Anarcestida Lower–Upper Devonian	Goniatitic; trilobate	At most one A^1 0–18 U,[1] sometimes split	Open to almost involute planispiral	31, o, p t–v, 33, i
Goniatitida Middle Devonian–Upper Permian	Goniatitic, some with slightly slit lobes; trilobate	1 to several A, 0 to several U, E sometimes split	Closed planispiral, frequently involute	31, w, x 33, f–h
Clymeniida Upper Devonian	Goniatitic; trilobate	Few sutural elements, sometimes dorsal saddles	Closed planispiral, usually evolute, siphon internal	31, y
Prolecanitida Upper Devonian–Upper Triassic	Goniatitic, some with slightly slit lobes; trilobate	A not developed, 2–7 U, E sometimes slit	Closed planispiral	33, u 35, a
Ceratitida Lower Permian–Upper Triassic	Ceratitic to ammonitic; quadrilobate	Unipolar laceration; A absent, only U developed	Mostly closed planispiral, some heteromorphous forms, sometimes distinct sculpturing	33, e 35, b–h
Ammonitida Lower Jurassic–Upper Cretaceous	Ammonitic, quinquelobate	Bipolar laceration; A absent, only U developed	Mostly closed planispiral, frequently heavily sculptured	32 33, b, k–p 34 35, l–v
Phylloceratida Triassic–Upper Cretaceous	Ammonitic, quadri- and quinquelobate	Many narrow lobes; multiple, symmetrical splits of one U leading to a sutural lobus; I^1 not lacerated	Frequently involute, only slightly sculptured shell	33, c 35 i, k
Lytoceratida Lower Jurassic–Upper Cretaceous	Ammonitic; quinquelobate	Few, wide lobes; mostly just 2 U	Frequently evolute, only slightly sculptured shell	33, d 36, a
Ancyloceratida Upper Jurassic–Upper Cretaceous	Ammonitic quadrilobate	Morphogenesis of the suture variable; usually few elements	Shell frequently aberrant (no closed spiral)	33, a 36, b–g

[1]Abbreviations: A = adventitious lobe, E = dorsal lobe, I = ventral lobe, U = umbilical lobe

FIGURE 36. Important representatives of the Ammonoidea. **a** Order Lytoceratida. **b–g** Order Ancyloceratida. **a** *Lytoceras*, 0.2x, Jurassic–Cretaceous; **b** *Ancyloceras*, 0.04x, Lower Cretaceous; **c** *Crioceratites*, 0.08x, Lower Cretaceous; **d** *Scaphites*, 0.2x, Upper Cretaceous; **e** Fragment of *Baculites* phragmocone, 0.8x; **f** *Baculites*, 0.08x, Upper Cretaceous; **g** *Turrilites*, 0.25x, Upper Cretaceous (after *Moore, Rieber*).

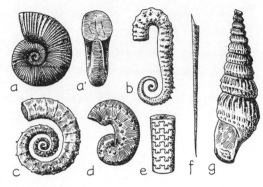

Ammonoidea most likely could have been derived directly or indirectly from the Phylloceratida.

There are three clearly recognizable phases of the Ammonoidea evolution. The first interval of prosperity from the Devonian to the Permian was followed by a severe reduction in abundance during the latest period of the Permian. The ammonoids flourished again during the Triassic, owing to the great variety of ceratitid forms. The crisis at the end of the Triassic was weathered by only one or at most a few evolutionary lines. The Jurassic and Cretaceous ushered in another great expansion of the Ammonoidea, followed by their extinction at the end of the Cretaceous.

Endocochliate Cephalopoda (Dibranchiata or Coleoidea)

The shell is contained within the soft body and is mostly an elongated, shoehorn-shaped structure with a narrow conical posterior. It is composed of either calcium carbonate or calcareous conchiolin, or of conchiolin only. Many forms have a greatly reduced shell. The mouth is surrounded by eight or ten arms equipped with hooklets or with suctorial disks. Strong mandibles, a radula, and an ink sac, typical organs of this group, were documented for several fossil representatives as well.

Order Belemnitida

(figures 37, 38, table 11)

Rostrum, with or without phragmocone, is usually the only part of the Belemnitida that is preserved. It consists of radially organized calcite prisms. In the massive distal part of the rostrum (the rostrum solidum), the prisms are arrayed such that they diverge toward the outside from an apical line. Besides, the rostrum is built of concentric

TABLE 11.   Overview of Orders of Endocochliate Cephalopoda

Order Geological Range	Number of Arms, Their Armament	Ink Sac	Inner Skeleton	Figure(s)
Sepioidea Upper Cretaceous– Recent	10, suctorial disks	Often present	Dorsally located sepium (cuttle-bone)	—
Teuthoidea Devonian–Recent	10, suctorial disks	Often present, also fossil documented	Keratinous gladius, in fossil forms sometimes calcified	38, i
Octobrachia Tertiary–Recent	8, suctorial disks	Frequently present	Small sepium, or completely reduced	—
Phragmoteuthida Permian– Upper Jurassic	10, double rows of hooklets	Documented	Conical phragmocone and a calcareous proostracum of three parts	38, a
Aulacocerida Carboniferous– Lower Jurassic	? 10, ? hooklets	?	Pointed conical phragmocone ensconced in a calcareous sheath (telum)	38, b, c
Belemnitida Lower Carboniferous, Lower Jurassic– Upper Cretaceous	10, hooklets	Documented	Sharp conical phragmocone ensconced in a massive calcitic rostrum, and a thin one-part proostracum	37 38, d–h

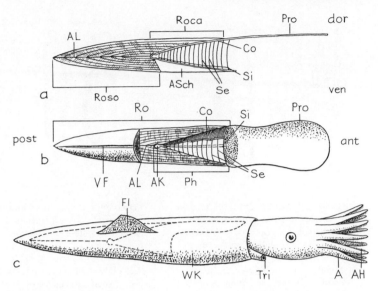

FIGURE 37. Order Belemnitida. **a** Schematic longitudinal section of the skeleton; **b** Ventral view of the skeleton, anterior portion partially removed; **c** Restoration of a living belemnite. A, arm, AH, hooklets, AK, protoconch, AL, apical line, ant, anterior, ASch, alveolar slit, Co, conotheca, dor, dorsal, Fl, lateral fin, Ph, phragmocone, post posterior, Pro, proostracum, Ro, rostrum, Roca, rostrum cavum, Roso, rostrum solidum, Se, septum, Si, siphon, Tri, hyponome (funnel), ven, ventral, VF, ventral furrow, WK, soft body.

layers, owing to the addition of material from the outside. These concentric layers are frequently alternatively lighter and darker. Ventrally, an alveolar slit may be formed in the region of the rostrum cavum. The rostra are usually roundish in cross section. Many representatives had developed ventral or apical furrows.

Bilaterally symmetrical vascular impressions on the upper surface and lateral double furrows may be seen on a well-preserved rostrum. The double furrows are considered to be impressions of cartilagineous side fins.

The posterior of the rostrum of some Belemnitida is extended into a long, tubular, loosely built epirostrum.

The conotheca (shell of the phragmocone) forms, dorsally, a very delicate, long proostracum. Septa of the phragmocone are close together, and the ventral siphuncle usually looks like a string of pearls.

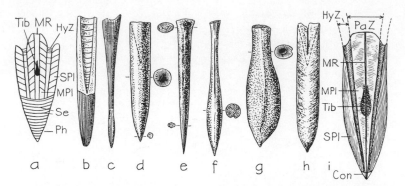

FIGURE 38. Representatives of the subclass Coleoidea. **a** Order Phragmoteuthida. **b, c** Order Aulacocerida. **d–h** Order Belemnitida. **i** Order Teuthoidea. **a** Skeleton of *Phragmoteuthis*, 0.5x, Middle to Upper Triassic; **b** *Aulacoceras*, 0.3x, Triassic, telum is opened anteriorly to show the phragmocone; **c** *Dictyoconites*, 3x, Permian–Triassic; **d** *Passaloteuthis*, 0.3x, Lower Jurassic; **e** *Megateuthis*, 0.08x, Middle Jurassic; **f** *Hibolithes*, 0.4x, Upper Jurassic–Lower Cretaceous; **g** *Duvalia*, 0.5x, Middle Jurassic–Lower Cretaceous; **h** *Belemnitella*, 0.3x, Upper Cretaceous; **i** Gladius of *Loligosepia*, 0.2x, Lower Jurassic. Con, conus, Hyz, hyperbolic zone, MR, middle rib, MPl, middle plate, PaZ, parabolic zone, Ph, phragmocone, Se, septum, SPl, lateral plate, Tib, ink sac (after *Naef, Rieber*).

Belemnites were probably fast, powerful, and constant swimmers, comparable to the modern Teuthoidea. Numerous belemnites, particularly those of the Cretaceous era, proved to be good index fossils.

Mollusca incertae sedis (figure 39)

Various tubular calcareous fossils are found in the Paleozoic deposits. The taxonomic position of the organism that yielded these hard parts has not yet been clarified. Because they are comparable to the molluscan shells, they are often considered to be related to the mollusks and therefore are mentioned in this connection here.

6. Class Tentaculitoidea (Cricoconarida)

Ordovician–Upper Devonian (figure 39, a)

This class includes shells that are small (2–30 mm), narrowly conical (apical angle 7–15°), straight or slightly curved, mostly cross-ribbed, and round in cross section. Some forms have developed only few septa. Because of the structure of the shell wall, the Tentaculitoidea are usually classified with or near the mollusks. However, other groups (e.g., Tentaculata) ought to be considered as well. Important genera are *Tentaculites, Nowakia,* and *Styliolina;* during the Silurian and Devonian they sometimes occurred in great abundance. The thin-shelled forms were probably planktonic, the thick-shelled forms vagile-benthic.

7. Class Calyptoptomatida

Cambrian–Permian (figure 39, c, d)

This class includes conical, up to 15-cm-long shells that are rounded, triangular, or oval in cross section. Most developed an operculum for closing the aperture. *Hyolithes,* the best-known form, shows, in addition to the operculum, two bent calcareous rods that are considered to be fin props. Recently, consideration has been given to assigning the "hyoliths" to their own extinct phylum, Hyolitha, close to the Mollusca, or classifying them with Annelida.

Group of Phyla Articulata

Animals with originally segmented body. Most have specialized organs for fast locomotion (segmented legs, wings). The chitinous cuticle that is the body covering of most members of this group is an efficient protective cover, and on the inside it affords attachment structures for striated muscles. Although they were originally inhabitants of the sea, many modern Articulata have successfully adapted to life on dry land and in the air.

Phylum Annelida

Cambrian–Recent

The Annelida are wormlike Articulata with usually metamerically segmented bodies. Their epidermis is protected by a protein-polysaccharide cuticle, and the body segments of many forms carry parapodia, stumpy locomotion organs. Chitinous setae occur frequently. Annelida yield few fossils, with the exception of representa-

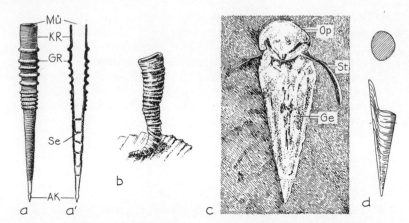

FIGURE 39. Fossils of uncertain taxonomic status. **a** Class Tentaculitoidea: *Tentaculites,* 3x, Lower Silurian–Upper Devonian; **b** Cornulitidae (family of uncertain status): *Cornulites,* 3x, Middle Ordovician–Middle Devonian; **c, d** Class Calyptoptomatida: **c** *Hyolithes,* 2x, Middle Cambrian, **d** Restoration and cross section of a *Hyolithes* shell, 1x. AK, protoconch, Ge, shell, GR, large annulus, KR, small annulus, Mü, aperture, Op, operculum, Se, septum, St, calcareous spine (after *Moore*).

tives of the family Serpulidae, which inhabit permanent calcareous tubes.

Family Serpulidae (feather-duster worms)

Cambrian–Recent (figure 40)

Serpulidae are heteronomously segmented, sessile-benthic Annelida, carrying many setae (class **Polychaeta**). On the peristomium (body section containing the mouth), they carry a funnel-shaped ring of tentacles that serves for straining and filtering food particles from the water. Both dorsal tentacles can form a keratinous or calcareous operculum. The calcareous dwelling tubes are from a few millimeters to over 10 cm long, usually irregularly bent. Many tubes show several external longitudinal keels or stripes in addition to the always prominent transverse growth lines. Externally similar gastropode shells are easily distinguishable from the serpulid dwelling tubes by the different shell structure.

Representatives of the class **Myzostomoida**, 3–5 mm long parasites of the echinoderms, have occurred since the Silurian; they are documented from cysts found on the Crinoidea.

FIGURE 40. Phylum Annelida,
class Polychaeta, family
Serpulidae. **a** *Serpula*,
0.5x, Recent; **b** *Spir-
orbis*, 10x, Pliocene;
c *Rotularia*, 1x, Upper
Cretaceous–Eocene;
d *Sarcinella*, 1x, Jurassic–Cretaceous. AS,
growth lines, KaR, cal-
careous tube, Op,
operculum, Ten, tentacle
(after *Hennig, Moore*).

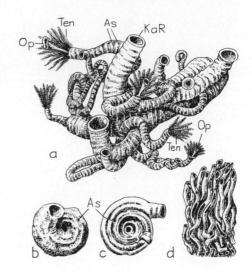

Scolecodonts

Upper Cambrian–Upper Cretaceous (figure 41)

Chitinous, occasionally calcified jaw elements of fossil Polychaeta are
termed scolecodonts. The size of the jaw elements, which are similar
to those of recent Polychaeta, ranges from 0.1 to approximately
5 mm. The scolecodonts, usually dark and shiny, are found in dark
schists and limestones, mostly as isolated elements, seldom as more
complete maxillary apparatus. Most common are scolecodonts of the
Ordovician to Devonian periods.

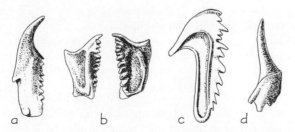

FIGURE 41. Scolecodonts. **a** *Arabellites*,
20x, Ordovician–Devonian; **b** *Pal-
eoenonites*, from both sides, 25x,
Ordovician; **c** *Leodicites*, 35x,
Ordovician–Silurian; **d** *Diopatraites*,
9x, Ordovician–Devonian (after
Moore).

Phylum Onychophora

The phylum Onychophora includes animals that combine characteristics of both the Annelida and Arthropoda. They are documented
as fossils only through two genera, that is, *Aysheaia* from the Middle
Cambrian Burgess Shales, and *Xenusion* from a Nordic sandstone
boulder. The sandstone could even be as old as late Algonkian.
Aysheaia is comparable to the modern *Peripatus*.

Phylum Arthropoda

Lower Cambrian–Recent (figures 42–51)

Arthropoda, currently represented by more than 1 million living species, are by far the species-richest animal phylum. Several taxonomic
groups adapted to terrestrial habitats, and insects even took to the air.
Their body plan is closely related to that of the Annelida, though of
course adult arthropods do not exhibit internal metamerism. The body
and the jointed appendages are covered by a multilayer cuticle. Chitinous parts of many arthropods may incorporate calcium carbonate and
calcium phosphate or sometimes also hard organic substances, thus
becoming stiffened, sclerotized. Thanks to such mineral deposition,
arthropod remains are frequently preserved as fossils. The chitinous
cuticle of the insects provides inside structures for the attachment of
muscles and organs and affords efficient protection. The original
segmentation is visible on the surface of the cuticle when the segments have not been fused into larger sections. The individual parts
of the sclerotized cuticle of both the main body and the tubular
appendages are connected by joints that allow movement. The chitinous cuticle of a body section is not stiff all around; it is interrupted
by membranous zones. Thus, most arthropods have a dorsal part (tergite) flexibly joined with a ventral sternite. Some arthropods have, in
addition, a weakly sclerotized side plate (pleurite). The sclerotized
cuticle cannot grow along with the soft body and is therefore periodically shed. During this molt (ecdysis) the shell is torn along
unsclerotized ecdysial seams.

The arthropod body is usually divided into several distinct sections (head, thorax, and abdomen), each consisting of several segments. The head and parts of the thorax may be to a greater or lesser
degree fused into a cephalothorax, or prosoma.

Paired ventral appendages of the body are variably differentiated
and may be absent, particularly in the abdominal region.

It is assumed that the original extremities consisted of a main axis with two either movable or fixed lateral processes (outer exopodite and inner endopodite). The proximal exopodites may have developed into gills and/or swimming organs (natatorial legs), the main axes usually into walking legs, and the endopodites in the head section into masticatory organs (chewing lobes). Division of the Arthropoda into subphyla and classes is based on the differentiation of the (originally at least six) pairs of extremities of the head area into sensory organs (antennae) and mouthparts (table 12).

Subphylum Trilobitomorpha

(figures 42–46)

The Trilobitomorpha were predominantly marine arthropods with biramous legs of the trilobitan type (main axis as a walking leg and proximal endopodite as a gill) and a preoral pair of antennae. They occurred from the Lower Cambrian to the Upper Permian.

In addition to the voluminous class of the Trilobita, this subphylum encompasses several smaller classes of organisms found mostly in the Middle Cambrian **Burgess Shales** of British Columbia. These forms (figure 42, a–f) exhibit characteristics that bring to mind Merostomata and Crustacea, in addition to the typical characteristics of the Trilobitomorpha.

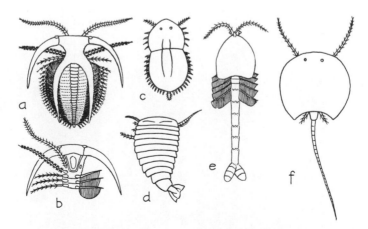

FIGURE 42. Subphylum Trilobitomorpha from the Middle Cambrian Burgess Shale. **a, b** Class Marellomorpha. **c, d** Class Merostomoidea. **e, f** Class Pseudonotostraca. **a, b** *Marrella*, 1.5x, **a** Dorsal view, **b** Ventral view; **c** *Naraoia*, 2x; **d** *Sidneyia*, 0.2x; **e** *Waptia*, 0.8x; **f** *Burgessia*, 2.2x (after *Moore*).

TABLE 12. Anterior Appendages and Body Division of Several Groups of Arthropoda

Subphylum: Class:	Trilobitomorpha	Chelicerata	Mandibulata Crustacea	Mandibulata Myriapoda and Insecta
Acron = Prostomium	—	—	—	—
Preantennal segment	—	—	—	—
1st antennal segment	Pair of antennae	—	Pair of antennae	Pair of antennae
2nd antennal segment	Biramous legs	Chelicerae	Pair of antennae	—
Mandibular segment	Biramous legs	Pedipalps	Mandibles	Mandibles
1st maxillar segment	Biramous legs	Pair of legs	Maxillae	Maxillae
2nd maxillar segment	Biramous legs	Pair of legs	Maxillae	Maxillae = lower lip
1st body segment	Biramous legs	Pair of legs	Pair of legs	Pair of legs
2nd body segment	Biramous legs	Pair of legs	Pair of legs	Pair of legs

Subphylum: Class:	Trilobitomorpha Trilobita	Chelicerata Merostomata	Arachnida	Mandibulata Crustacea	Insecta
Body division:	Cephalon	Prosoma = Cephalothorax	Prosoma	Head	Head
	Thorax	Opisthosoma = Abdomen { Mesosoma = Preabdomen / Metasoma = Postabdomen }	Abdomen	Thorax = Pereion / Abdomen = Pleon } Cephalothorax	Thorax { Prothorax / Mesothorax / Metathorax }
	Pygidium	Telson		Telson ≙ last abdominal segment	Abdomen

Class Trilobita

Cambrian–Permian (figures 43–46)

Marine Trilobitomorpha with a trilobate dorsal shield (figure 43, a) consisting of a cephalon (head shield), thorax, and pygidium (tail shield). Cephalon, thoracic segments, and pygidium were flexibly joined to each other. Most trilobites were thus able to roll up with head and tail touching.

In contrast to the heavily sclerotized tergites that form the dorsal armor, the ventral sternite and the extremities are not sclerotized (or are only lightly sclerotized) and are, therefore, only rarely fossilized. Trilobites molted several times during their life, so that one animal could provide numerous fossil exuviae.

CEPHALON (figure 43, b–o)

The cephalon consists of a median part (glabella) and two lateral cheeks. On the cheeks are compound eyes, either holochroal (with all ommatidia underneath a common cornea) or schizochroal (each of the relatively few ommatidia with its own cornea). Most trilobites show facial sutures, along which the cephalon splits during ecydysis (molt). The facial sutures separate the free cheek with an eye from the fixed cheek and palpebral lobe. Different shapes of the facial suture are recognized (e.g., proparian, opisthoparian, and gonatoparian). They serve taxonomic purposes. Cranidium is termed the central, undivided part of the cephalon (glabella and fixed cheeks). The cephalon is a fused product of at least five segments. Its ventral part carries one pair of antennae and three or four pairs of biramous legs. The furrows that frequently occur on the glabella are another indication of the original segmentation of the cephalon. Two unpaired sclerotized plates—the rostral plate and the hypostome—are usually attached to the frontal part of the cephalon. The mouth opening is directly behind the hypostome.

THORAX (figure 44)

The thoracic part of the dorsal shield consists of flexibly connected tergites, each tergite having a central axial ring and two lateral pleura (singular pleuron). The pleura are usually somewhat indented and frequently have a pleural furrow. Each thoracic segment carries ventrally a pair of biramous legs.

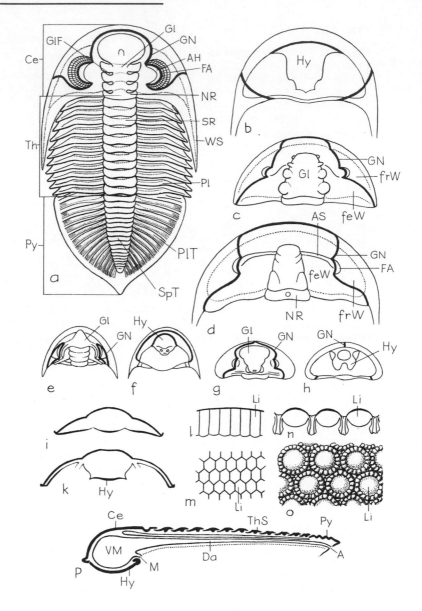

PYGIDIUM

Pygidium is a product of fusion of a smaller (originally) or greater (secondarily) number of segments. It consists of a central axial part and lateral pleural lobes. Ventrally there are biramous legs on each segment except the last one.

ONTOGENY (figure 45)

In the course of their ontogeny, trilobites undergo several larval stages, subsequently shedding their skins. The protaspid stage lasts from hatching at approximately 0.25 mm in size to the development of a protopygidium at about 1 mm. During the subsequent meraspid period, one segment is added after each ecdysis, beginning with the meraspid stage 0. At stage 0 protopygidium flexibly joins the cephalon. The final number of segments is attained at the holaspid period. During this time there is a six- to twelvefold increase in size. Further molts during the holaspid period merely enable further growth. Many trilobites may go through thirty or more ecdyses during their ontogeny.

CLASSIFICATION (figure 46)

The present classification of the Trilobita must not be considered definitive, in spite of being based on many different characteristics. New finds of fossilized soft-part remains raise some doubts even about the unity of the class.

FIGURE 43. Morphology of Trilobita. a Dorsal shield of *Odontochile* (schematic drawing), with a proparian facial suture; b Ventral aspect of *Odontochile* cephalon; c Cephalon of *Flexicalymene* with a gonatoparian facial suture; d Cephalon of *Ptychoporia* with an opisthoparian facial suture; e, f Dorsal and ventral views of *Kjerulfia* cephalon, with a metaparian facial suture; i, k Schematic cross sections of *Asaphus* cephalon; l–o Schematic sections of the surface of trilobitan eyes, l, m holochroal, n, o schizochroal; p Longitudinal section of a trilobite (schematic). A, anus, AH, palpebral lobe, AS, ocular ridge, Ce, cephalon, Da, intestine, FA, compound eye, feW, fixed cheeks, frW, free cheeks, Gl, glabella, GlF, glabellar furrow, GN, facial suture, Hy, hypostome, Li, lens, M, mouth, NR, occipital ring, Pl, pleuron, PlT, pleural lobe, Py, pygidium, SpT, axial lobe, SR, axial ring, Th, thorax, ThS, thoracic segment, VM, prestomach, WS, genal spine (after *Hupé, Moore*).

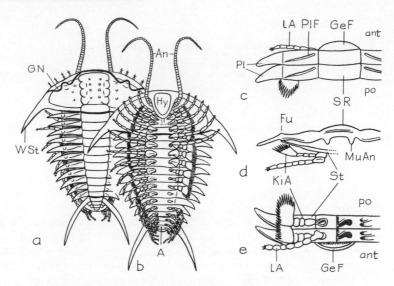

FIGURE 44. Class Trilobita. **a, b** Restoration of *Ceraurus pleurexanthemus,* 1x, Middle Ordovician; **c–e** Thoracic segments and biramous legs (schematic). A, anus, An, antennae, ant, anterior, Fu, fulcrum, GeF, joint surface, GN, facial suture, Hy, hypostome, KiA, gill branch, La, walking leg, MuAn, muscle attachment area, Pl, pleura, PlF, pleural furrow, po, posterior, SR, axial ring, St, sternite, WSt, genal spine (after *Moore*).

1. Order Agnostida

Lower Cambrian–Upper Ordovician (figure 46, e)

Very small forms with two or three thoracic segments, mostly without eyes. Cephalon and pygidium are similar in size and appearance. There are no facial sutures. Agnostida, which are very important for the division of Upper Cambrian, had to be separated from the remaining Trilobita on the basis of newly discovered soft-part remains.

2. Order Redlichiida

Lower to Middle Cambrian (figure 46, a)

Opisthoparian, relatively large trilobites, frequently with genal spines (spines on the cheeks), with many thoracic segments and small pygidium.

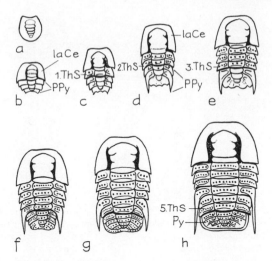

FIGURE 45. Class Trilobita, growth stages of *Shumardia pusilla,* Lower Ordovician. **a** Protaspis, 25x; **b–g** Meraspis stages, **b** Meraspis stage 0, 25x, **c** Meraspis stage 1, 22x, **d** Meraspis stage 2, 24x, **e** Meraspis stage 3, 26x, **f** Meraspis stage 4, 20x, **g** Meraspis stage 5, 18x; **h** Holaspis, 15x. laCe, larval cephalon, PPy, protopygidium, Py, pygidium, ThS, thoracic segment (after *Moore*).

3. Order Corynexochida

   Cambrian

   Trilobita similar to the Redlichiida, but with larger pygidium and fewer thoracic segments.

4. Order Ptychopariida

   Lower Cambrian–Permian (figure 46, b, c, g)

   The order with the greatest number of forms, predominantly opisthoparian. Glabella usually exhibits distinct, backward-directed lateral furrows.

5. Order Phacopida

   Lower Ordovician–Upper Devonian (figure 46, d, f, h)

   Post-Cambrian trilobites with mostly proparian or gonatoparian facial sutures and an anteriorly broadened glabella. The compound eyes are

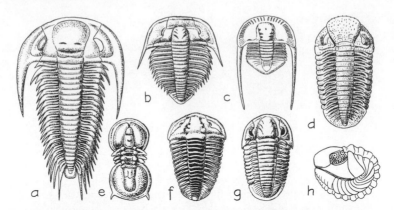

FIGURE 46. Class Trilobita. a *Paradoxides,* 0.3x, Middle Cambrian; b *Olenus,* 2x, Upper Cambrian; c *Trinucleus,* 1x, Ordovician; d *Phacops,* 0.5x, Silurian–Devonian; e *Agnostus,* 2.5x, Upper Cambrian; f *Calymene,* 0.3x, Ordovician–Devonian; g *Proetus,* 1x, Ordovician–Devonian; h *Phacops* rolled up, side view, 1.5x, Middle Devonian (after *Hupé, Moore*).

either holochroal or schizochroal; the thorax consists of eight to nineteen segments.

6. Order Lichida

Lower Ordovician–Upper Devonian

Medium-sized to very large (70 cm long) trilobites with an opisthoparian facial suture and variously developed glabella.

7. Order Odontopleurida

Middle Cambrian–Upper Devonian

Small, spiny trilobites with opisthoparian facial sutures.

ECOLOGY

Trilobita lived mostly on the bottom of well-aerated seas. Blind forms probably rooted in the sediments. Fossilized traces of trilobites, moving and at rest, are preserved in many sediments. The action of the many extremities is recognizable in the braidlike *Cruziana* traces left by forms plowing through the sediments. Trilobites probably fed on small remnants of plants and animals.

The geographic distribution of trilobites makes it possible to distinguish various faunal provinces, particularly for the Cambrian.

Subphylum Chelicerata

Lower Cambrian–Recent (figure 47)

Arthropods with a body consisting of prosoma, opisthosoma, and telson (postanal spine, tail spine) (cf. table 12). There are six pairs of appendages on the prosoma. The foremost segment of the opisthosoma (which originally had twelve segments) is usually reduced. Segments of the anterior part of opisthosoma, the mesosoma, may carry appendages; those of the posterior part, the metasoma, usually do not. The telson, when developed, is dagger- or oar-shaped. The first of the six pairs of extremities on the prosoma is modified into chelicerae; the second pair, frequently into pedipalps; and the remaining pairs, into walking legs.

1. Class Merostomata

Lower Cambrian–Recent (figure 47, a–l)

Aquatic, gill-breathing Chelicerata with a long opisthosoma and a strong telson. The second pair of extremities is developed into walking legs; the sixth pair, often into natatorial legs.

Subclass Xiphosura (horseshoe crabs)

Lower Cambrian–Recent (figure 47, a–g)

This subclass encompasses marine Merostomata with a long telson. Most members are relatively small (5–7 cm) and, apart from the telson, compact. These ancient chelicerates are still represented by the recent genus *Limulus,* which may grow as large as 60 cm.

Subclass Eurypterida (giant sea scorpions)

Ordovician–Permian (figure 47, i–l)

Small to very large (up to 2 m) Merostomata with a long stretched-out body. Opisthosoma consists of twelve segments, of which meso-

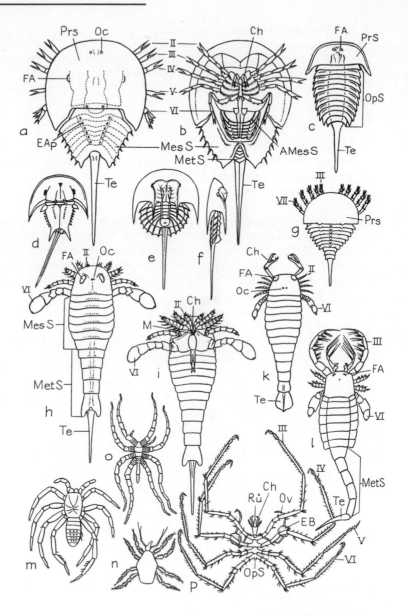

CLASS PANTOPODA 107

soma accounts for seven and metasoma for five. The stratigraphically older Eurypterida lived in the sea; among the younger ones brackish- and freshwater forms are also found.

2. Class Arachnida

Silurian–Recent (figure 47, m, n)

Tiny (0.1 mm) to medium-sized (approx. 20 cm), air-breathing Chelicerata, predominantly terrestrial. There are many parasitic forms in addition to the more usual carnivorous ones. Prosoma has six pairs of extremities: one pair each of chelicerae and pedipalps, and four pairs of walking legs. Opisthosoma is seldom segmented; instead, it has developed into a soft-skinned sac, with extremities either absent or present only as transformed rudiments. Telson is absent. Among the many orders, for example, Scorpionida (scorpions), Araneida (spiders), and Acarida (mites), there are but few well-documented fossil forms. Many of those found are fossilized in amber.

3. Class Pantopoda (sea spiders)

Lower Devonian–Recent (figure 47, o, p)

Marine Chelicerata with a thin prosoma, long walking legs, and a tiny opisthosoma. In addition to the chelicerae, pedipalps, and four pairs of walking legs, there is a single pair of ovigerous (egg-carrying) legs. The position of the Pantopoda in the classification

FIGURE 47. Subphylum Chelicerata. a–g Class Merostomata, Order Xiphosura. h–l Class Merostomata, Order Eurypterida. m Class Arachnida, Order Araneida. n Class Arachnida, Order Acarida. o, p Class Pantopoda. a, b Limulus, 0.2x, Recent; c Aglaspis, 0.4x, Upper Cambrian; d Psammolimulus, 0.6x, Lower Triassic; e, f Euproops, 0.75x, Lower Devonian; g Weinbergina, 0.25x, Lower Devonian; h, i Eurypterus, 0.25x, Silurian; k Pterygotus, 0.2x, Lower Devonian; l Mixopterus, 0.06x, Lower Devonian; m Protolycosa, 1.5x, Carboniferous; n Bdella, 15x, Oligocene; o Palaeopantopus, 0.3x, Lower Devonian; p Nymphon, 2.5x, Recent. AMesS, appendages of the mesosoma (phyllopodia), Ch, chelicerae, EAp, entapophyses, EB, egg mass, FA, compound eye, M, mouth, MesS, mesosoma, MetS, metasoma, Oc, ocellus/median eye, OpS, opisthosoma, Ov, oviger/egg carrier, Prs, prosoma, Rü, proboscis, Te, telson, I–VII appendages of the prosoma (after Hennig, Moore).

system is variously determined, depending in part on whether their chelicerae and pedipalps are homologous with those of other chelicerates.

Subphylum Mandibulata

Lower Cambrian–Recent (figures 48–51)

Aquatic or terrestrial arthropods with one or two pairs of antennae.

Class Crustacea

Lower Cambrian–Recent (figures 48–50)

Predominantly aquatic, gill-breathing Mandibulata. Differences in the morphology of fully developed adults of the various groups may be great; however, a unifying characteristic of the class is the nauplius larva, which is very similar in all groups.

The crustacean body may be divided into head, thorax, and abdomen (cf. table 12). The head is formed by fusion of the prostomium with probably six segments. Numbers of thoracic and abdominal segments vary between five and sixty. The head and thorax of many crustaceans are fused into a cephalothorax. Characteristic is the carapace, a shell-like dorsal doubling of the body wall originating on the second maxillar segment. It is originally bivalved; secondarily it may consist of one piece. The outer layer of the carapace is often heavily sclerotized. The carapace may either cover some of the frontal parts of the body or enclose the body entirely. The paired extremities, though quite differently formed in the different regions of the body, can be traced to the biramous leg.

CLASSIFICATION

The multitude of Crustacea is divided into nine subclasses; however, only Branchiopoda, Ostracoda, Cirripedia, and Malacostraca are of paleontological importance.

Conchostraca (figure 48, c–h) belong to an order of the subclass Branchiopoda, of which the recent representatives live mainly in fresh water. Conchostracan genera, *Leaia* and *Cyzicus (Isaura),* are frequently fossilized in freshwater deposits. Their body is enclosed in a bivalval carapace that shows concentric growth lines.

Subclass Ostracoda

Cambrian–Recent (figure 48, i–m)

Ostracods are small (usually 0.3–5 mm, rarely up to 40 mm) crustaceans with a bivalval, mussellike carapace (shell) that envelops the entire body. Most have a heavily calcified outer layer of the carapace and thus are easily preserved. The edge of the inner carapace layer of many ostracods is also heavily calcified. The two halves of the carapace are unequal in size and are connected by an elastic dorsal ligament. The ligament serves to open the valves; the transverse adductor muscles close them. The structure of the muscle attachment areas is of taxonomic importance. Many forms achieve a better closure, having developed a hinge with teeth and sockets along the dorsal side. There is a transparent area (ocular spot) in the shell, over the laterally located compound eyes. The valves are porous and are often sculptured in various degrees. Many ostracods are distinctly sexually dimorphic. The nonstretchable integument, of which the carapace is the most prominent part, is shed several times during the growth period.

Ostracoda are primarily marine and are excellent index fossils. As most of them are benthic, they can also serve as facies fossils.

Subclass Cirripedia (barnacles)

Cambrian–Recent (figure 49, a–l)

Cirripedia are generally small, either sessile or parasitic. Sessility is an extremely rare trait among crustaceans.

In the sessile, predominantly marine Cirripedia, the frontal, preoral part of the head is formed into a stalk (peduncle) or into an attachment plate. The body is enveloped in a mantle that grows out of the carapace fold. Calcareous plates that provide protection are secreted by the outer part of the mantle. The relationship of barnacles to the Crustacea is revealed primarily by the occurrence of the typical crustacean larva—nauplius.

Members of the order Thoracica are of paleontological significance: goose barnacles (Lepadomorpha) and acorn barnacles (Balanomorpha). Lepadomorpha (Cambrian–Recent) are generally 5–10 cm but rarely up to 1 m long, are attached to substrate by a peduncle, and are protected by a varying number of calcareous plates. Balanomorpha (Upper Cretaceous–Recent) attach directly to the substrate by a plate that is usually calcified. The calcareous plates that

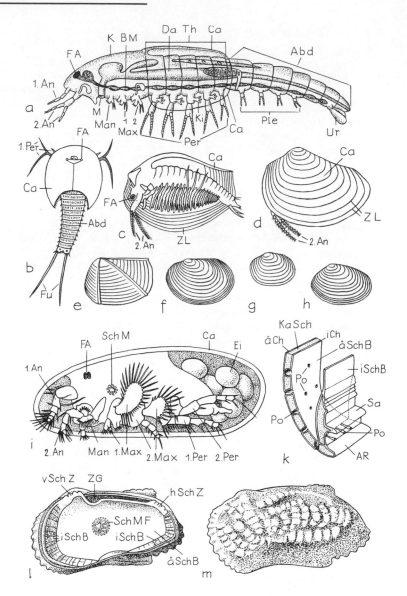

are embedded in the mantle overlap, forming a resilient wall. Their number varies (originally there were eight), and their structure is specialized, with inside tubes. The opening in the wall can be closed by an operculum consisting of two pairs of plates (terga and scuta).

Lepadomorpha are most common in shallow waters and on firm substrates. The nauplius larva attaches to the substrate in such fashion that, after metamorphosis, the carinal side faces up (geotropic growth).

Subclass Malacostraca

Cambrian–Recent (figures 49, m–s, 50)

The malacostracan thorax consists of eight segments, the abdomen of seven or six segments. Abdominal legs (pleopods), which are usually present, show a different structure than the thoracic legs (pereiopods). Most Malacostraca live in the sea, some in freshwater, and numerous Isopoda are terrestrial.

The enormous variety of forms in this subclass is classified into superorders and orders on the basis of development of the extremities, the cephalothorax and the carapace, as well as on the basis of the number of abdominal segments and of some other characteristics.

There are abundant fossil remains, primarily of the order **Decapoda** (Lower Triassic–Recent). The first three of the six pereiopods of the Decapoda are developed into maxillipeds, the other five into walking legs; the fourth thoracic pair usually carries powerful claws.

FIGURE 48. Crustacea. a Schematic of a generalized crustacean body organization. b–h Subclass Branchiopoda: b Order Notostraca: *Triops*, 1.5x, Upper Triassic. c–h Order Conchostraca: c *Limnadia*, 3x, Recent; d *Cyzicus* (= *Isaura*), 7x, Recent; e *Leaia*, 6x, Upper Carboniferous; f–h *Palaeestheria*, 3x, Upper Triassic. i–m Class Ostracoda: i Schematic of the structure; k Section of the shell margin; l, m *Cythereis*, 50x, Lower Cretaceous; l Right valve from the inside, m Left valve from the outside. Abd, abdomen, äCh, outer chitinous layer, An, antenna, AR, outer margin, äSchB, outer shell wall, BM, abdominal nervous cord, Ca, carapace, Da, intestine, Ei, eggs, Fa, compound eye, Fu, furca, hSchZ, posterior hinge tooth, iCh, inner chitinous layer, iSchB, inner shell wall, K, head, KaSch, calcified layer, Ki, gills, M, mouth, Man, mandible, Max, maxilla, Per, pereiopods/thoracopods, Ple, pleopods, Po, pore, Sa, seam, SchM, adductor muscles, SchMF, adductor muscle attachment area, Th, thorax, Ur, uropods, vSchZ, front hinge tooth, ZG, dental socket, ZL, growth lines (after *Hennig, Kühn, Moore, Moore* and *Teichert, Sars, Triebel, Warth*).

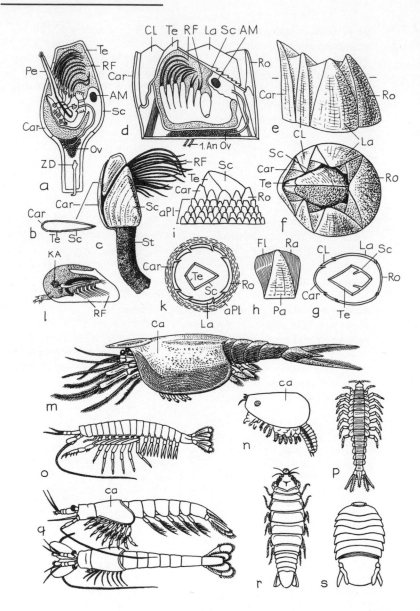

In addition to the long-tailed decapods with strongly developed abdomens and tail fans, short-tailed forms with reduced abdomens and no distinct tail fans have occurred since the Upper Lias. The abdomen is either curved in (hermit crabs) or flattened against the cephalothorax (Brachyura/crabs). Short-tailed forms have repeatedly evolved from long-tailed ones, that is, they are polyphyletic.

Class Myriapoda

Middle Devonian–Recent (figure 51, a, b)

Myriapoda are terrestrial Mandibulata with tracheas. Many are excellently preserved in amber.

Class Insecta (Hexapoda)

Devonian–Recent (figure 51, f–l)

The insect body consists of three sections: head, thorax, and abdomen. The thorax has three segments and carries three pairs of walking legs. Winged insects develop two pairs of wings, one on the second and one on the third thoracic segments. Being primarily terrestrial, insects have a tracheal system. Mouthparts are modified according to the various specialized food habits; they can be developed into organs for biting, piercing, or sucking. Although modern

FIGURE 49. Class Crustacea. a–l Subclass Cirripedia. m–s Subclass Malacostraca. a–c *Lepas*, 0.5x, Pliocene–Recent, a Schematic of the structure, b Side view, c Cross section; d–h *Balanus*, Eocene–Recent, d Schematic of the structure, e Side view of the wall, 2x, f Wall and operculum from above, g Cross section of the wall, h Wall plate; i, k *Catophragmus*, 1x, Recent, i Side view, k Schematic of the plate arrangement; l Cypris larva of *Lepas*, 14x; m *Nahecaris* (superorder Phyllocarida), 0.8x, Lower Devonian; n *Canadaspis* (superorder Phyllocarida), 0.7x, Middle Cambrian; o *Uronectes* (superorder Syncarida), 1.2x, Permian; p *Pleurocaris* (superorder Syncarida), 2.5x, Upper Carboniferous; q *Schimperella* (superorder Peracarida), 2.5x, Lower Triassic; r *Palaega* (superorder Peracarida, order Isopoda), 2x, Lias; s *Eosphaeroma*, a fossil isopod, 1.4x, Tertiary. AM, adductor muscles, An, antennae, aPl, accessory plates, Ca, carapace, Car, carina, CL, carinolateral, Fl, wing, KA, compound eye, La, lateral, Ov, ovary, Pa, parietal, Pe, penis, Ra, radius, Ro, rostrum, RF, pereiopods, Sc, scutum, St, stalk, Te, tergum, ZD, cement gland (after *Dechaseaux, Henning, Moore* and *Teichert, Müller, Riedl*).

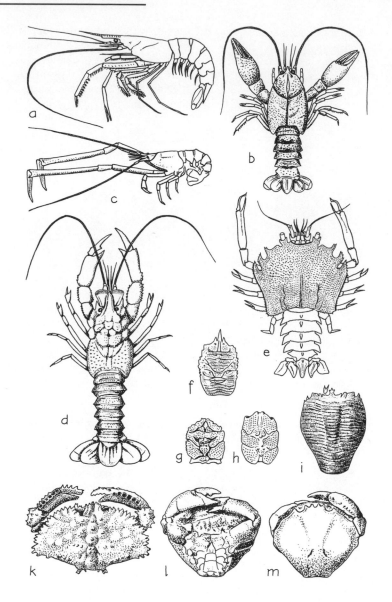

insects account for three-quarters of all animal species, they are of little paleontological importance. Most are small, terrestrial animals with a chitinous integument that is poorly suited to fossilization. As a consequence they are rarely preserved, although they are sometimes perfectly preserved in amber, a fossilized resin of pines and other trees. Well preserved insects have been found in several fine-grained sediments as well.

The original insects (subclass **Apterygota**) were wingless. The oldest ones known are from Middle Devonian. Several superorders of the winged insects (subclass **Pterygota**) have been documented since the Upper Devonian; they are distinguished mainly on the basis of the development of wings and mouthparts.

Palaeodictyopteroidea (figure 51, g) was a superorder of large insects (up to 50 cm) with horizontally outstretched wings that did not fold. They have been documented from the Carboniferous and Permian periods.

Superorder **Odonatopteroida** (dragonflies sensu lato) are insects with biting mouthparts and folding wings that are attached to the thorax by a narrow base. They have been documented since the Upper Carboniferous. *Meganeura* from the Upper Carboniferous attained a maximal wingspread of 70 cm. Typical dragonflies (figure 51, i) are found in the Upper Jurassic lithographic limestones of Solnhofen.

Arthropoda incertae sedis (figure 51, c–e)

There are numerous fossil arthropods, paleozoic ones in particular, whose systematic status is still disputed.

The remains of a giant, freshwater or perhaps also terrestrial arthropod, *Arthropleura* (figure 51, c, d), are known from the Upper Carboniferous. The animal had many segments, much like the myriapods, yet its extremities were biramous legs of the trilobitan type.

FIGURE 50. Class Crustacea, Superorder Eucarida, Order Decapoda. **a–e** Long-tailed decapods. **f–m** Short-tailed decapods (Brachyura). **a** *Aeger*, 0.3x, Upper Triassic–Jurassic; **b** *Eryma*, 1x, Jurassic–Lower Cretaceous; **c** *Mecochirus*, 0.4x, Jurassic; **d** *Pemphix*, 0.7x, Middle Triassic; **e** *Eryon*, 0.35x, Middle Jurassic–Lower Cretaceous. **f** Carapace of *Galathea*, 2x, Upper Cretaceous–Recent; **g** Carapace of *Prosopon*, 0.7x, Upper Jurassic; **h** Carapace of *Pithonoton*, 0.8x, Upper Jurassic; **i** Carapace of *Ranina*, 0.25x, Old Tertiary; **k** *Lobocarcinus*, carapace and claw of the first pereiopod pair, 0.3x, Eocene; **l, m** *Harpactoxanthopsis*, ventral and dorsal view, 0.25x, Eocene (after *Dechaseaux, Moore, Zittel*).

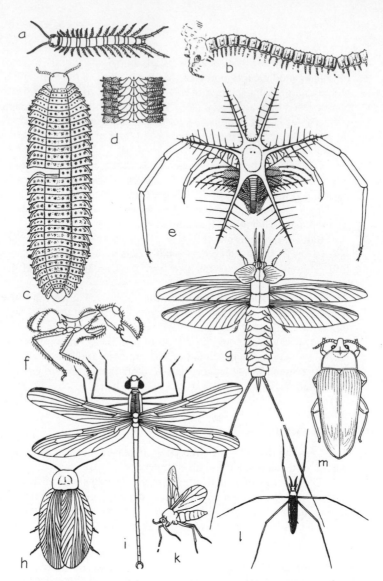

FIGURE 51. Phylum Arthropoda. **a, b** Class Myriapoda; **c, d** Arthropoda incertae sedis; **e–m** Class Insecta. **a** *Lithobius*, 2x, Lower Oligocene (amber); **b** *Euphoberia*, 0.7x, Upper Carboniferous; **c, d** *Arthropleura*, 0.08x (**c** dorsal view, **d** section of the ventral side with extremities), Upper Carboniferous; **e** *Mimetaster*, 1x, Lower Devonian; **f** *Gaesomyrmex* (ant), 5x, Miocene; **g** *Stenodictya*, 0.4x, Upper Carboniferous; **h** *Aphthoroblattina* (cockroach), 0.6x, Upper Carboniferous; **i** *Tarsophlebia* (dragonfly), 0.9x, Upper Jurassic; **k** *Plegiofungivora* (gnat), 5x, Jurassic; **l** *Chresmoda*, 0.4x, Upper Jurassic; **m** *Rhinohelaeites* (beetle), 1.7x, Eocene (after *Laurentiaux, Moore, Müller, Stürmer*).

Mimetaster (figure 51 e), a marine arthropod from the Lower
Devonian, is probably related to either trilobites or chelicerates.

Phylum Echinodermata

Lower Cambrian–Recent (figures 52–61)

Most recent Echinodermata are pentamerously radially symmetrical.
Because they are lacking a distinct head or front end of the body, it is
customary to refer to the oral (with mouth) side and the opposite
aboral (= apical) side when describing them. All recent echinoderms
have a water-vascular system (ambulacral system), with a circumoral
ring and five radial canals with ambulacral tentacles. The various
types of echinoderm habitus are to a large extent defined by the structure and orientation of the ambulacral system. The echinoderm
skeleton consists of calcareous plates, or ossicles, either rigidly or
flexibly connected together and externally covered by the epidermis.
The skeletal elements are porous calcareous crystals of mesodermal
origin; the volume of the pores may account for up to 60% of the
total volume of the ossicle. In a living animal, the pores are filled by
the stroma, a syncytium that secretes and, when necessary, rebuilds
the calcareous lattice (stereome). Such a structure affords great
strength with minimal weight. During fossilization the stroma is frequently replaced by oriented calcite, resulting in massive calcspar
crystals. However, on a polished thin section, it is usually possible to
discern the original microstructure. Thanks to the calcareous skeletal
elements, echinoderms are among the more common fossils.
    All echinoderms are marine, predominantly stenohaline. Post-
larval stages are benthic, either vagile or sessile; the larvae, however,
are planktonic. The stratigraphically oldest echinoderm remains come
from the Lower Cambrian. Their greatest expansion occurred during
the Paleozoic.

CLASSIFICATION

Formerly, two subphyla were recognized, as defined by mode of
living and body structure: Pelmatozoa (primarily sessile) and
Eleutherozoa (vagile). Today, however, the great diversity of
echinoderm forms is being classified in six subphyla, with a total of
twenty-one classes. But the systematic status of some of the twenty-
one classes is still disputed. The recent echinoderms belong to only
four classes (Crinoidea, Stelleroidea, Echinoidea, Holothuroidea).

Subphylum Homalozoa

Middle Cambrian–Devonian

Paleozoic echinoderms with an asymmetric, usually flattened body, consisting of theca and stalk. The theca is made up of numerous

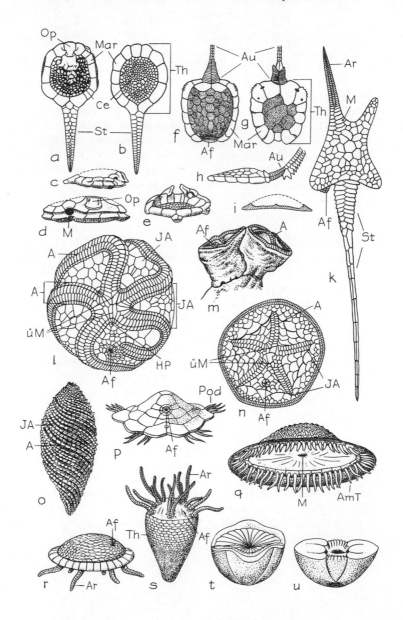

polygonal plates that may be differentiated into strong marginal plates (marginalia) and inner plates (centralia). There are three classes: Homostelea, Stylophora, and Homoiostelea. Homalozoans probably lived loose on the sea bottom and could anchor themselves by their stalks.

Subphylum Blastozoa

Middle Cambrian–Permian

Paleozoic echinoderms with a saclike or spherical theca with brachioles (short, unbranched arms). The theca usually attaches to the substrate by a stalk, rarely directly. The oral side of the body, with the ambulacral system, is oriented upward. The theca is built of numerous polygonal ossicles and, in the more highly developed forms, is usually pentamerously symmetrical.

Class Eocrinoidea

Middle Cambrian–Silurian (figure 53, o–q)

Eocrinoidea encompass the more rare Blastozoa that are not, or are only imperfectly, radially symmetrical and have characteristic pores along the plate margins.

FIGURE 52. Older Paleozoic Echinodermata. a–e Class Homostelea. f–l Class Stylophora. k Class Homoiostelea. l–n Class Edrioasteroidea. o Class Helicoplacoidea. p Class Ophiocistioidea. g Class Cyclocystoidea. r Class Camptostromatoidea. s Class Lepidocystoidea. t, u Echinodermata incertae sedis. a–e Trochocystites, 0.8x, Middle Cambrian, a from above, b from below, c side view, d oral, 1x, e aboral. f–i Mytrocystites, 0.8x, Middle–Upper Ordovician, f from above, g from below, h side view, i cross section of the theca. k Dendrocystoides, side view, 0.6x, Upper Ordovician; l Edrioaster, oral side, 0.9x, Middle Ordovician; m Cyathocystis, side view, 1.5x, Middle Ordovician; n Hemicystites, oral side, 1.4x, Ordovician; o Helicoplacus, side view, 1x, Lower Cambrian; p Volchovia, aboral side, 0.5x, Lower Ordovician; q Cyclocystoides, side view (oral side probably pointing downward), 1.4x, Middle Ordovician; r Camptostroma, side view (aboral side up) 0.5x, Lower Cambrian; s Lepidocystis, side view, 0.4x, Lower Cambrian; t Cymbionites, side view, 2x, Middle Cambrian; u Peridionites, side view, 3x, Middle Cambrian. A, ambulacral field, Af, anus, AmT, ambulacral tentacle, Ar, arm, Au, aulacophore, Ce, centralia, HP, hydropore, IA, interambulacrum, M, mouth, Mar, marginalia, Op, operculum, Pod, podia, St, stem, Th, theca, üM, covered mouth (after Moore, Nichols).

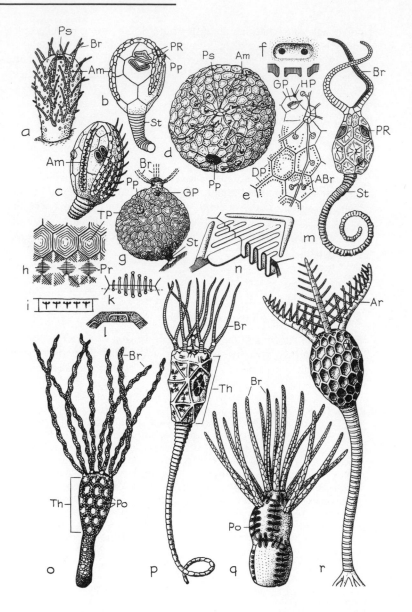

Class Cystoidea

Ordovician–Devonian (figure 53, a–n)

Thecas of cystoidea are saclike to spherical and may be composed of more than 2,000 or as little as 13 polygonal plates. The stalk is frequently short or absent.

Of special taxonomic value for the Cystoidea are pores and various pore rhombes in the thecal plates. There may be many simple haplopores or many diplopores or numerous or even individual pore rhombes. The pores and pore rhombes are partially covered by a thin calcareous layer; they probably had a respiratory function.

Ambulacral grooves number from two to five and may be either simple or branched. Originally short, they were secondarily lengthened. They lead from the peristome to the distal attachment areas for brachioles. On the oral side of the theca next to the peristome are the periproct (anal field), hydropore, and gonopore.

Class Blastoidea

Silurian–Permian (figure 54)

The rigid, bud-shaped theca of most Blastoidea (their most successful period was Lower Carboniferous) is regular, pentamerously symmetrical, and carried on a stem with roots. The peristome is centrally located on top. Attachment areas for the numerous brachioles and pores leading into hydrospires are found along the edges of the five

FIGURE 53. Paleozoic Echinodermata. a–n Class Cystoidea. o–q Class Eocrinoidea. r Class Paracrinoidea. a *Fungocystites*, restoration, 1.3x, Middle Ordovician; b *Lovenicystis*, 2x, Silurian; c *Callocystis*, 0.7x, Middle Silurian; d *Glyptosphaerites*, 1x, Ordovician; e Part of the oral side of *Glyptosphaerites*, 2x; f Schematic of a diplopore, view from the outside (upper) and in section (below); g–l *Echinosphaerites*, 0.5x, Ordovician; g Theca, 1x; h Part of the theca, with the outlines of thecal plates (above showing the outermost, sculptured layer, below with the layer removed to show the pore rhombes); i–l Schematic of the pore rhombe structure (cross section, and sections parallel and perpendicular to the plate margin); m *Pleurocystites*, 0.5x, Middle Ordovician; n Schematic of one pore rhombe; o *Gogia*, 1.2x, Middle Cambrian; p *Macrocystella*, 1.2x, Lower Ordovician; q *Lichenoides*, 1.6x, Middle Cambrian; r *Comarocystites*, 0.3x, Middle Ordovician. ABr, attachment field for a brachiole, Am, ambulacral field, Ar, arm, Br, brachioles, DP, diplopore, GP, gonopore, HP, hydropore, Po, pore, Pp, periproct, PR, pore rhombe, Ps, peristome, St, column, Th, theca, TP, thecal plate (after *Moore, Nichols*).

ambulacral fields. The hydrospires are thin calcareous lamellae, each
with a fold. The hollow of this fold opens to the outside via an orally
located spiraculum. Five or ten spiracula surround the peristome.
Hydrospires are thought to have served a respiratory function. The
periproct is located either apart, between the two rear ambulacral

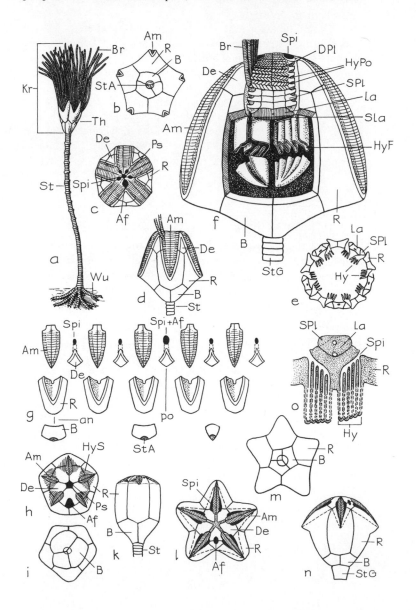

fields, or is incorporated into the rear spiraculum. During phylogenesis the number of ambulacral fields increased and the hydrospire folds multiplied. The theca, originally elongated, became shorter and in many forms was secondarily bilaterally symmetrical.

Subphylum Crinozoa

Cambrian–Recent (figures 53, r, 55, 56)

Crinozoa are small to very large (up to 20 m long) echinoderms. Their bodies, usually pentamerously symmetrical, consist of a theca with an upward-oriented oral side and laterally attached, upward pointing arms. A stem (column) is usually present. Most Crinozoa, in particular the fossil ones, are sessile, attached to substrate either by the stem or by the base of the cup. The first free-moving forms appeared during the Mesozoic. The greatest abundance of crinozoan forms has been recorded for the Silurian and Permian periods.

Class Paracrinoidea

Middle Ordovician (figure 53, r)

Paracrinoidea encompass nonpentamerous Crinozoa with, at most, four arms that are equipped with side branches (pinnulae). The theca consists of many irregularly arranged plates. Paracrinoidea are a relatively homogeneous group.

FIGURE 54. Class Blastoidea. a Restoration of *Orophorocrinus*, 0.5x, Lower Carboniferous. b–g *Pentremites*, Lower Carboniferous. a–e Theca in an aboral, oral, and side view, and in cross section, 1.5x; f Schematic of the structure of the theca (partial section), 4x; g Details of the theca; h–k Theca of *Codaster*, oral, aboral, and side views, 1x, Lower Carboniferous; l–n Theca of *Orophorocrinus*, oral, aboral, and side views, 0.8x, Lower Carboniferous. o Part of the cross section of *Orophorocrinus* theca, showing the position of spiracula and hydrospires. Af, anus, Am, ambulacral field, an, anterior, B, basals, Br, brachioles, De, deltoids, DPl, cover plates of the ambulacral grooves, ambulacra, Hy, hydrospires, HyF, hydrospire fold, HyPo, hydrospire pores, HyS, hydrospire slit, Kr, crown of theca and brachioles, La, lancet plate, po, posterior, Ps, peristome, R, radials, SLa, under lancet plate, Spi, spiraculum, SPl, side plates, St, column, StA, column attachment field, StG, columnal (stem segment), Th, theca or calyx/cup, Wu, root (after *Moore, Nichols*)

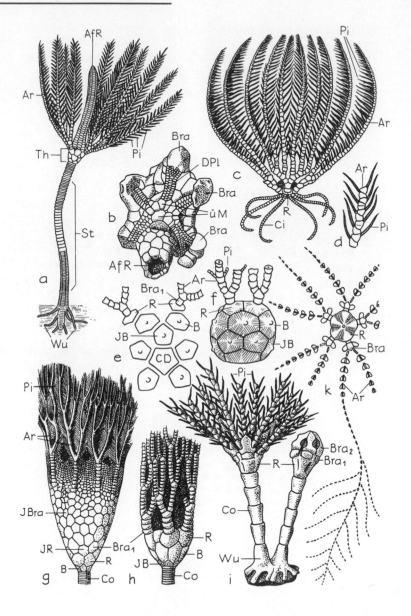

Class Crinoidea (sea lilies and feather stars)

Lower Ordovician–Recent (figures 55, 56)

Of all the echinoderm classes, the Crinoidea contain the greatest variety of forms. They are stalked or secondarily stalkless, and most are distinctly pentamerously symmetrical. The body of the pentamerously symmetrical Crinoidea consists of a theca with five simple or branched arms and a segmented stem or column, with either a root or an attachment disk. The ambulacral system running through the radially attached arms represents a substantial enlargement of the food-gathering system.

The structure of the theca is of primary taxonomical importance. The theca consists of a (usually rigid) dorsal cup and its roof, the tegmen. The cup may be monocyclic or dicyclic. Above the uppermost segment of the stem, the monocyclic cup has first a ring of five basals, then in alternating positions another ring of five radials. The first brachials (segments of the arms) are attached to the radials. In the case of a dicyclic base, another ring of alternating plates, infrabasals, is found between the last column segment and the basals. In many forms the bases of arms with their branches may be incorporated in a relatively large cup.

The thecal cover, or tegmen, is either a vaulted calcareous wall or a leathery skin armored with calcareous plates. The peristome is located in its center, the periproct laterally behind. Forms occurring in still waters may have a long, conical anal tube. Ambulacral grooves (ambulacra) are impressed into the tegmen and protected by cover plates. The grooves overlap onto the arms, which are built from the brachials as either uniserial or biserial. Fine side branchlets, called pinnulae, occur in most forms. Brachialia and pinnulae are flexibly joined together by connective tissue and muscle fibers and

FIGURE 55. Class Crinoidea. a, b, h Subclass Inadunata. g Subclass Camerata. c–f, i, k Subclass Articulata. a *Dictenocrinus,* 0.6x, Silurian; b Tegmen of *Cyathocrinites,* 1.5x, Carboniferous; c *Neometra* (a stemless crinoid), 0.9x, Recent; d *Neometra,* part of an arm, 1.6x; e, f *Marsupites,* Cretaceous; e Details of theca; f Side view of the theca, 0.3x; g *Scyphocrinites,* 0.25x, Silurian–Lower Devonian; h *Cyathocrinites,* 0.6x, Silurian; i *Eugeniacrinites* (restoration), 1.5x, Upper Jurassic; k *Saccocoma,* 2x, Upper Jurassic. AfR, anal tube, Ar, arm, B, basal, Bra, brachial, CD, centrodorsal, Ci, cirri, Co, columnal, DPl, cover plates of the ambulacral grooves, ambulacra, IB, infrabasal, IBra, interbrachial, IR, interradial, Pi, pinnulae, R, radial, St, stalk, Th, theca (cup), üM, covered mouth, Wu root (after *Bather, Hyman, Meek* and *Worthen, Ubaghs*).

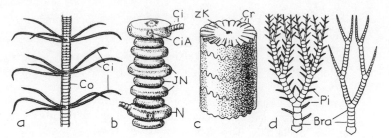

FIGURE 56. Class Crinoidea. Structure of the stem and arms. **a** *Teliocrinus,* 1x, Recent; **b** Schematic of the stem structure; **c** *Encrinus,* 2x, Middle Triassic; **d** Schematic of branching of the arms. Bra, brachial, Ci, cirrus, CiA, cirrus attachment field, Co, columnal (stem segment), Cr, crenulae, IN, internodal, N, nodal, Pi, pinnulae, zK, central canal (after *Moore*).

are, together with the entire skeleton, covered by the epidermis. The column varies in length, from short to very long (up to 20 m). It is built partly of movable and partly of rigidly connected segments (columnals). Most columnals are either round or pentamerously symmetrical. On their upper and lower surfaces are fine ridges (crenulae) and grooves (crenellae), arranged in pentamerous patterns that are specific to individual genera. They serve as joints and for the attachment of connective tissue and musculature. Whorls of cirri, which have anchoring and respiratory functions, are often attached to the columnals.

During times of their greatest prosperity, from the Silurian to the Permian, stalked Crinoidea preferred shallow areas of the sea. In contrast, the few recent Crinoidea inhabit deeper water, and the stemless feather stars prefer the littoral zone.

Remains of Crinoidea are very common fossils. However, the segmented skeletons are seldom connected because they fall apart when the connective tissue decays. Being originally quite light, the skeletal parts are often transported for various distances before they are finally deposited. Echinoderm breccia consist to a large extent of disjointed skeletal segments of crinoids, particularly of their stems and arms.

TAXONOMY AND DISTRIBUTION

The taxonomy of Crinoidea is based on the structure of the dorsal cup (rigid or flexible), calyx base (mono or dicyclic), tegmen (rigid or flexible, with or without an anal tube), arms (branched or undivided, uni- or biserial, with or without pinnulae), and stalk and on several further characteristics. The following four subclasses are distinguished:

Camerata (about 2,500 species, Ordovician–Permian)
(figure 55, g),
Inadunata (about 1,750 species, Ordovician–Triassic)
(figure 53, a, b, h),
Flexibilia (about 300 species, Middle Ordovician–
Upper Permian), and
Articulata (about 500 species, Triassic–Recent)
(figure 55, c–f, i, k).

In addition to many stalked forms, stalkless ones such as the recent feather stars and the *Saccocoma* (figure 55, k), a Jurassic floating form, belong to the Articulata. The oldest floating Crinoidea are known from early Upper Triassic. *Seirocrinus* from the Upper Lias with its 20-m-long column is the longest representative of Invertebrata.

Subphylum Asterozoa

Lower Ordovician–Recent (figures 57, 58)

Echinodermata with a flattened, star- or disk-shaped body consisting of a central disk and radially arranged arms that can be either very short or quite long. The oral side with the ambulacral system is oriented downward. The skeleton is built of flexibly connected plates, vertebral ossicles, and spines.
    Skeletal segments of decayed Asterozoa are frequent in many sediments; rarely, however, are they found still connected.
    The subphylum has only one class (**Stelleroidea**), which unites three subclasses: Somasteroidea, Asteroidea, and Ophiuroidea.

Subclass Somasteroidea

Lower Ordovician–Recent (figure 57, o–s)

Stelleroidea with a large central disk that carries wide-based arms. In the middle of the oral side, the skeleton of the arms has a double row of ambulacral plates from which side rays formed by virgalia diverge. They support a taut leathery skin. On the aboral side, the arms are protected by small, star-shaped calcareous particles. Besides Lower Ordovician forms, only one recent genus, *Platasterias*, is known.

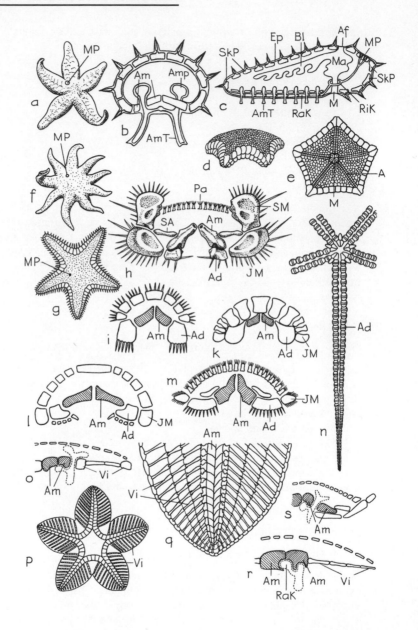

Subclass Asteroidea (sea stars, starfish)

Lower Ordovician–Recent (figure 57, a–n)

Stelleroidea whose arms attach to the central disk by means of a wide base. Blind sacs of the intestine lead into hollow arms built of various skeletal plates. Many of those carry movable spines. Radial canals of the ambulacral system run orally outside the skeleton. Fine small canals branch from them to the ambulacral tentacles whose ampullae are located inside the skeleton. The anus lies aborally. Numerous fossil genera and species have been distinguished, based on the skeletal structure of the central disk and arms and on the shape of the skeletal particles.

Subclass Ophiuroidea (brittle stars, serpent stars)

Lower Ordovician–Recent (figure 58)

Stelleroidea whose arms attach to the central disk by means of a narrow base. The arm skeleton consists of central vertebral ossicles (arm vertebrae) and lateral, dorsal, and ventral shield plates (lateral, dorsal, and ventral shield). The lateral shields often carry tubercles that in turn carry spines. The vertebrae, which are jointed and connected by muscle fibers, developed phylogenetically from two ambulacrals embedded in the center of the arm. In the original Ophiuroidea, they

FIGURE 57. Subphylum Asterozoa.
a–n Subclass Asteroidea. o–s Subclass Somasteroidea. a *Asterias,* 0.15x, Recent; b Schematic cross section of the arm; c Schematic of the body structure; on the right the section leads through the central disk, on the left through an arm; d, e *Metopaster* lateral aboral and oral (from below) views, 0.3x, Cretaceous; f *Solaster,* 0.15x, Recent; g *Asteropecten,* 0.25x, Recent; h–m Schematics of the arm structure; h *Asteropecten,* Recent, i *Palasterina,* Silurian; k *Urasterella,* Middle Ordovician–Permian; l *Petraster,* Lower Ordovician; m *Platanaster,* Middle Ordovician; n *Urasterella,* 1.25x, Upper Ordovician, oral side with strong adambulacrals; o *Archegonaster,* Lower Ordovician, cross section of an arm; p–r *Chinianaster,* Lower Ordovician, p Restoration, oral side, q Structure of the arms, oral side, 2.5x, r Cross section of an arm; s *Platasterias,* Recent, cross section of an arm. A, ambulacral field, Ad, adambulacral, Af, anus, Am, ambulacral, Amp, ampulla, AmT, ambulacral tentacle, Bl, blind sac of the intestine, Ep, epidermis, IM, inferomarginal, M, mouth, Ma, stomach, MP, madreporite, Pa, paxillae, RaK, radial canal, RiK, ring canal, SA, superambulacral, SkP, skeletal plate, SM, submarginal, Vi, virgalia (after *Moore, Nichols*).

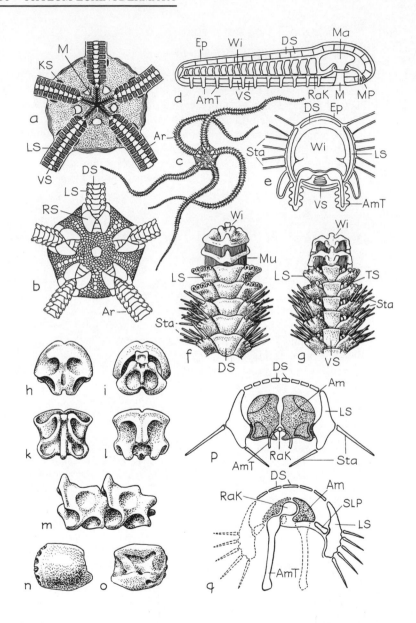

still consist of two separate halves. The vertebrae enable the serpent stars to move their arms in a "snakewise" fashion.

Many ophiuroid taxa are distinguished by the structure of the skeleton, particularly of the arms, and by the form of the skeletal plates.

Isolated skeletal parts of Ophiuroidea are frequent in many sediments and can be obtained by elutriation.

Subphylum Echinozoa

Lower Cambrian–Recent (figures 52, l–q, 59–61)

Pentamerous or bilaterally symmetrical echinoderms with spherical, disk-, heart-, or cucumber-shaped bodies without arms. Most forms are free moving.

Class Edrioasteroidea

Lower Cambrian–Recent (figure 52, l–n)

Echinozoa with a flexible, disk-, or sac-shaped theca; the peristome, periproct, hydropore, and ambulacral fields are arranged like a five-rayed star on the upper surface of the theca. The skeletal plates are originally embedded into and held together and covered by skin. Ambulacral fields are straight in the stratigraphically oldest forms and increasingly curved in the younger ones.

The Edrioasteroidea, most of which were sessile, are very rare.

FIGURE 58. Subclass Ophiuroidea.
a–c *Ophiura*, Recent, a Central disk, 1.25x, oral side, b Central disk, 1.25x, aboral side, c Aboral side, 0.5x. d Schematic of the body structure; on the right the section leads through the central disk, on the left through an arm; e Schematic cross section of an arm; f, g *Ophiocoma*, part of an arm, partially uncovered, 3.5x, Recent, f Aboral side, g Oral side; h–m Vertebral ossicle of *Hallaster*, 9x, Silurian, h Proximal, i Distal, k Aboral, l Oral views, m Side view of two ossicles; n, o *Ophioderma*, lateral shield, viewed from outside and inside, 12x, Upper Jurassic; p *Hallaster*, cross section of an arm (vertebral ossicle is in two parts), Middle Ordovician; q *Eophiura*, cross section of an arm (vertebral ossicle is in two parts), Lower Ordovician. Am, ambulacrals, AmT, ambulacral tentacles (podia), Ar, arm, DS, dorsal shield, Ep, epidermis, KS, bursal slit, LS, lateral shield, Ma, stomach, MP, madreporite, M, mouth, Mu, muscles, RaK, radial canal, RS radial shield, SLP, sublateral plate, Sta, spine, TS, tentacle scales, VS, ventral shield, Wi, vertebral ossicle (after *Hess, Moore, Nichols, Ubaghs*).

Class Echinoidea (sea urchins)

Ordovician–Recent (figures 59–60)

Pentamerous or bilaterally symmetrical echinozoans with a spherical, apple-, disk-, or heart-shaped theca. The skeletal plates of most forms have smooth edges and form a rigid corona. In a few forms they are movable, somewhat overlapping, and covered by a rough skin. Many skeletal plates carry tubercles (mamelons) with movable spines. The corona is composed of five ambulacral and five interambulacral fields (A-fields and IA fields), running like meridians, from the peristome (which always points downward) to the aboral apex. In the modern sea urchins, both A- and IA-fields consist of two rows of plates, with zigzag lines of contact. The ambulacrals have double pores for the podia. Each field begins with a single plate on the apex: A-fields with an ocular plate, IA-fields with a genital plate. One of the five genital plates also serves as a madreporite plate and thus as the beginning of the water vascular system. The periproct (anal field) is located either on the apex (regular Echinoidea) or in the posterior IA-field (irregular Echinoidea). Peristome and periproct are protected by a skin that is armored with calcareous platelets, leaving small openings for the mouth and anus. When a chewing apparatus, the Aristotle's lantern, is present, it is anchored with muscles and tendons to auricles and/or apophyses along the inner side of the periproct.

FIGURE 59. Class Echinoidea. **a, b** Aboral (**a**) and oral (**b**) views of a corona of a regular sea urchin (*Cidaris*). Spines are removed from most of the surface to show the structure underneath. On the right, only the plate margins are shown, not the mamelons or the ambulacral pores. **c** Schematic of body structure of a regular sea urchin (showing the section on the right in the plane of the ambulacral, on the left of the interambulacral field). **d** *Plegiocidaris* (interambulacral plate with its spine), 1x. **e** Tridactylous pedicellariae. **f, g** Cross section and plane view of a cidarian, in the region of the A and IA fields. **h, i** Inner margin of the peristome (perignathic girdle) of a cidarian (**h**) and a pedinid (**i**). **k–n** Cross and longitudinal sections of spines of *Strongylocentrotus* (**k, l**) and *Diadema* (**m, n**), 10x. **o, p** Chewing apparatus of *Strongylocentrotus* in side view and aborally, 1.5x Recent; **q** Pyramid with an aulodont tooth; **r** Pyramid with a stirodont tooth. A, ambulakralfield, Ac, acetabulum, Af, anus, Amp., ampulla, AmT, ambulacral tentacles (podia), AP, ambulacral plate, Apo, apophyses, Aur, auricle, Da, intestine, Go, gonad, GP, genital plate, IA, interambulacral field, IAP, interambulacral plate, Ki, pyramid (jaw), LG, ligamental fossa, M, mouth, MP, madreporite plate, Mu, muscles, OP, ocular plate, Pe, pedicellariae, PeK, pedicellar valve, pincer valve, PG, plate margin, Pp, periproct, Ps, peristome, RaK, radial canal, Sch, spine shaft, SHa, spinal neck, SK, milled ring, SS, secondary spine, Sta, spine, SW, secondary tubercles, W, mamelon, WHo, tubercle field, WK, tubercle head, WKe, tubercle base, Z, tooth (after *Hess, Hesse, Moore, Termier* and *Termier*).

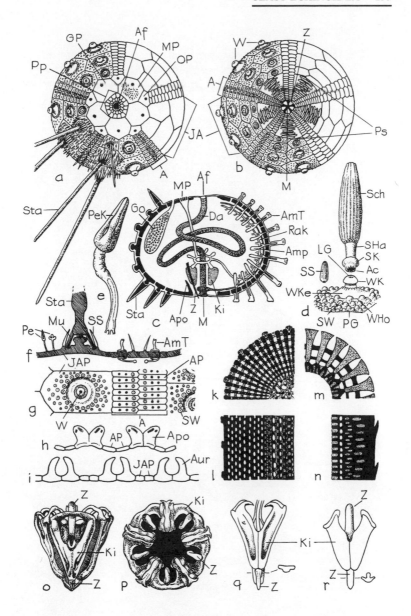

TAXONOMY

The following characteristics are of primary importance for the division of Echinoidea into subclasses, superorders, and orders: position of the periproct (in the apex = endocyclic, or outside of it = exo-

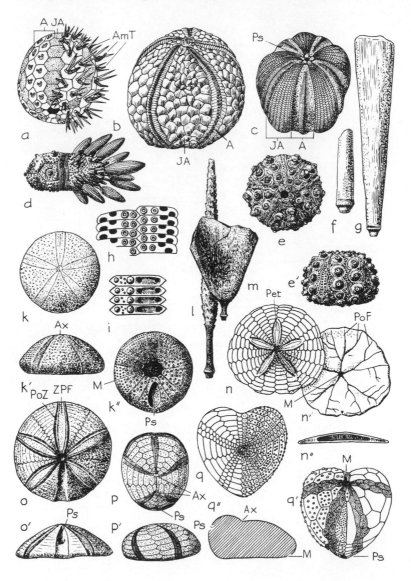

cyclic); type of corona (flexible or rigid); number of meridional rows of plates in the A and IA fields; type of ambulacral plates (simple and small, or compound and large); type of poral zones (oral and aboral or oral only, petaloid or not); presence or absence of chewing apparatus, tooth type, and structure of the perignathic girdle (auricles and apophyses); structure of the main spines; and presence or absence of fascioles (bands of finest tubercles) as well as their position and number.

Subclass Perischoechinoidea

Ordovician–Recent (figures 59, a, b, 60, a–d, h, i, l, m)

Perischoechinoidea are characterized by an endocyclic periproct, simple ambulacral plates, pore zones remaining the same from peristome to apex, and a round to pentagonal ambitus (the widest diameter perpendicular to the oral-aboral axis). The number of A and IA plate rows varies in the older forms; there are always two per field in the younger forms. Most of the younger forms have a rigid corona.

Subclass Euechinoidea

Carboniferous, Triassic–Recent (figure 60, e–g, k, n–q)

Euechinoidea encompass regular (endocyclic) and irregular (exocyclic) sea urchins. Most have a rigid corona. The regular

FIGURE 60. Class Echinoidea. a–d, h, i, l, m Subclass Perischoechinoidea. e–g, k, n–q Subclass Euechinoidea, e regular, k, n–q irregular. a *Bothriocidaris*, 1x, Upper Ordovician; b *Aulechinus*, 0.9x, Upper Ordovician; c *Melonechinus*, 0.35x, Lower Carboniferous; d *Plegiocidaris*, 0.5x, Upper Jurassic; e, e' *Hemicidaris*, 0.5x, Upper Jurassic, f, g Spines of *Hemicidaris*, 1x, Malm; h *Plegiocidaris*, part of an ambulacral field, 5x; i *Rhabdocidaris*, 4 ambulacral plates, 5x; k–k'' *Holectypus* (aboral, lateral, oral views), 0.7x, Middle and Upper Jurassic; l, m spines of *Rhabdocidaris*, 0.4x, Middle and Upper Jurassic; n–n'' *Scutella* (aboral and oral views and cross section), 0.4x, Miocene; o, o' *Clypeus* (aboral, rear view), 0.4x, Middle Jurassic; p, p' *Collyrites* (aboral, side view), 0.7x, Middle and Upper Jurassic; q–q'' *Micraster* (aboral and oral views, and longitudinal section), 0.5x, Upper Cretaceous. A, ambulacral field, AmT, ambulacral tentacles (podia), Ax, apex, M, mouth, Pet, petaloid, PoF, food grooves, PoZ, poral zone, Ps, peristome, ZPF, interporal field (after *Fischer, Hess, Moore, Zittel*).

Euechinoidea have megaplates in the A fields. The irregular Euechinoidea, with a more or less bilaterally symmetrical corona, may be divided into the superorders **Gnathostomata** (with the chewing apparatus) and **Atelostomata** (without one). Most of the Atelostomata are heart shaped and have only fine spines. Many of them have the three anterior A fields located separately from the two posterior ones on the aboral side. In such cases the apical shield is elongated. The peristome may be shifted forward on the oral side and have two lips. Irregular sea urchins appeared for the first time during the Lower Jurassic.

ECOLOGY

Regular sea urchins are exclusively epibenthic; irregular ones live in the sediment. The shallow-burrowing Gnathostomata keep their petaloid aboral parts clear of the sediment. In contrast, Atelostomata live completely buried in soft sediments; they move through it with the aid of their brushlike spines, leaving behind sediment-filled passages.

Class Helicoplacoidea

Lower Cambrian (figure 52, o)

Old Paleozoic Echinozoa with a theca of spirally arrayed, loosely connected plates and a spiral ambulacral system. Theca of the single known genus, *Helioplacus,* was probably elastic.

Class Ophiocistioidea

Lower Ordovician–Upper Silurian, ? Middle Devonian (figure 52, p)

A small group of old Paleozoic Echinozoa with pentamerous, low, helm-shaped thecas of large, firmly connected calcite plates. The ambulacral tentacles (podia), limited to the oral side (which is always oriented downward), are very large and in many forms are encrusted with calcite platelets.

Class Cyclocystoidea

Middle Ordovician–Middle Devonian (figure 52, q)

Small group of Echinozoa with a disk-shaped, probably flexible theca of many calcite platelets that were to some extent concentrically arrayed. Podia were probably arranged in a ring on the distal end of the ambulacral grooves of the oral side. The oral side of a live animal probably pointed up.

Class Holothuroidea (sea cucumbers)

? Cambrian, Ordovician–Recent (figure 61)

Echinozoa with an elongated, sac-shaped theca and laterally located mouth and anus (on opposite poles). The theca is formed by a leathery skin with loosely embedded sclerites—small, calcite bodies, often shaped like wheels, hooks, anchors, or plates. Many sclerites are species specific. The esophagus is encircled by a pentamerous calcareous ring (the dental ring) composed of alternating radial and interradial plates. Holothurian sclerites can be isolated from many sediments; they are of some importance for stratigraphy.

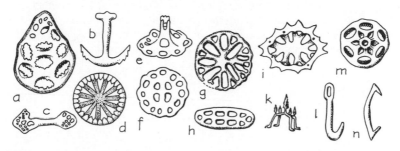

FIGURE 61. Sclerites of representatives of the class Holothuroidea, from different periods. Scale, 15–130x. **a, b** Tertiary, **c–f** Jurassic, **g–k** Triassic, **l–n** Paleozoic (after *Moore, Mostler*).

Echinodermata incertae sedis (figure 52, r–u)

Echinoderm remains that could not be conveniently placed into any of the known classes have been found in the Lower Cambrian strata. As a result, a separate class, **Camptostromatoidea,** was established for a radially symmetrical, medusalike form (*Camptostroma,* figure 52, r). Another class, **Lepidocystoidea,** was established for another free-living, polypous form (*Lepidocystis,* figure 52, s).

To date no class has been assigned to the cup-shaped skeletons of *Cymbionites* (figure 52, t) and *Peridionites* (figure 52, u); both were formed by only a few plates. The Lower Cambrian Echinodermata and some further older Paleozoic forms (not discussed here) as well as the Homalozoa, the Edrioasteroidea, and the Eocrinoidea witness the amazing variety and abundance of echinoderms that existed toward the beginning of the Cambrian period.

Phylum Hemichordata (Branchiotremata)

Middle Cambrian–Lower Carboniferous, Cretaceous–Recent (figures 62–63)

The phylum Hemichordata encompasses bilateral Coelomata, whose body has three parts: prosoma, mesosoma, and metasoma. The stomochord is everted from the dorsal section of the pharynx, and there are one or more pairs of gill slits. The presence of this branchial region of the intestine places Hemichordata near the chordates. Of two extant classes, **Enteropneusta** and **Pterobranchia,** both of which have numerous Recent marine representatives, undisputed fossil documentation exists only for the Pterobranchia.

Class Pterobranchia

Ordovician, Jurassic–Recent (figure 62, a, b)

Genus Rhabdopleura

Upper Cretaceous–Recent (figure 62, a, b)

Rhabdopleura, a recent colonial representative of Pterobranchia, is very important for comparison with the Graptolithina, which were limited to the Paleozoic. The tiny zooids of *Rhabdopleura* arise by budding from a stolon and live in unattached tubes, about 3 mm long. The proximal ends of the free tubes, as well as the ends of tubes attached to the substrate, consist of semicircular stripes connected by zigzag seams. The periderm tubes are composed of scleroproteins. The chitinous section of the stolon is designated pectocaulus; the non-chitinized section, gymnocaulus.

Class Graptolithina

Middle Cambrian–Lower Carboniferous (figures 62, c–s, 63, s)

Graptolithina are extinct colonial organisms that share many common characteristics of the skeletal structure with the Rhabdopleurida. A colony (rhabdosome) consists of a large number of small chambers (thecas) that issue from one another or from a conical first chamber (sicula) that arises by sexual reproduction.

The scleroprotein skeleton was originally flexible. Its wall, the periderm, consists of two layers: a thick, transparent inner layer of half-rings that are attached dorsally and ventrally in zigzag fashion, and the outer, usually opaque, laminar cortex.

A rhabdosome structure begins with a sicula that is composed of a prosicula and a metasicula. The first bud penetrates to the outside through a foramen of the sicula (caused by resorption) and builds a theca. Budding at the first and then the following thecas eventually results in the formation of a rhabdosome. The shape of the rhabdosome depends both on the type of budding and on the shape and arrangement of the thecas.

The sicula may carry ventrally a spine-shaped process, the virgella, and dorsally a pair of buccal spines. A filament, the nema, extends from the tip of the virgella. In graptolites where the thecas hang from the sicula, the nema remains free. However, when budding is directed toward it, the nema is incorporated into the dorsal area of the thecas or, when two rows are formed, into the middle lamella.

An incorporated nema is termed virgula. Graptolites have bipartite thecas, each with proximal protheca and distal metatheca. The consecutive prothecas form a common canal through which the stolon runs; the buds issue from the stolon. Neighboring metathecas share a common wall, termed interthecal septum.

ECOLOGY

Graptolites led in part a sessile and in part a planktonic life. The planktonic life was made possible by the formation of various floating organs (figure 62, p–s), such as an expanded, widened nema (nematularium) or a ramified virgella (virgellarium).

Some graptolites have been found in groups (synrhabdosomes) with numerous rhabdosomes hanging from a common swim bladder (pneumatophore). These synrhabdosomes also contained organs designated as gonothecas.

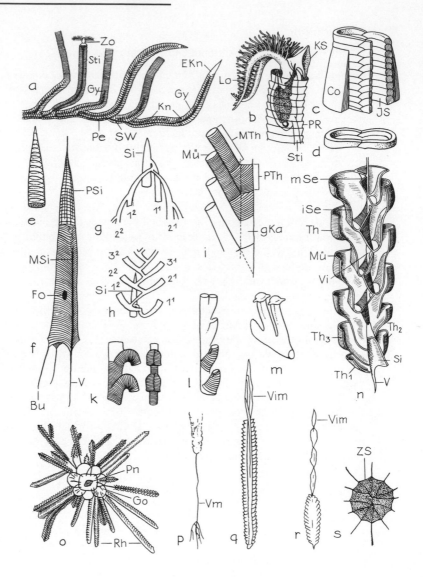

TAXONOMY

The class Graptolithina contains six orders, the most important being Dendroidea and Graptoloidea.

Order Dendroidea

Middle Cambrian–Lower Carboniferous (figure 63, a–d)

Heavily ramified Graptolithina with various thecas (auto-, bi-, and stolothecas) and chitinized stolons are placed in the order Dendroidea. Most representatives were sessile, but it must be assumed that *Dictyonema* was epiplanktonic. In this genus the individual branches of the rhabdosome were connected by dissepiments.

Order Graptoloidea

Lower Ordovician–Lower Devonian (figures 62, e–s, 63, e–z)

Graptoloidea are characterized by uniform thecas; however, their shape may have gradually changed during the development of the rhabdosome. Because the presence of stolons in this group has not been proven to date, it is likely that they were not chitinized. Graptoloidea most likely developed from Dendroidea, in particular from such forms as the *Dictyonema*.

FIGURE 62. Phylum Hemichordata. **a, b** Class Pterobranchia. **c–s** Class Graptolithina. **a** Section of a *Rhabdopleura* colony, 10x; **b** Zooid of *Rhabdopleura* inside the distal part of the periderm tube, 50x; **c, d** Structure of periderm of Graptolithina (schematic); **e** Prosicula, an early stage with a spiral line, 40x; **f** Sicula, 25x; **g, h** Various arrangements of thecas (schematically); **i** Structure of a uniserial rhabdosome; **k–m** Various types of thecas, about 6x; **n** Structure of a biserial rhabdosome of *Climacograptus*, 16x; **o** Synrhabdosome of *Glossograptus* (from above), 0.6x, Ordovician; **p** *Climacograptus* with a virgellarium, 5x, Lower Silurian; **q, r** Rhabdosome with a nematularium, **q** *Cystograptus*, 0.8x, Lower Silurian; **r** *Petalograptus*, 1.6x, Lower Silurian; **s** *Loganograptus*, with a central disk, 0.7x. Bu, buccal spines, Co, cortex, EKn, end bud, Fo, foramen, gKa, common canal, Go, gonotheca, Gy, gymnocaulus, IS, inner layer of half-rings, iSe, interthecal septum, Kn, bud, KS, head shield, Lo, lophophore with tentacles, mSe, median longitudinal septum, MSi, metasicula, MTh, metatheca, Mü, aperture, Pe, pectocaulus, Pn, pneumatophore, PR, peridermal ring, PSi, prosicula, PTh, prototheca, Rh, rhabdosome, Si, sicula, Sti, stolon, SW, septum, Th, theca, V, virgella, Vi, virgula, Vim, virgularium, Vm, virgellarium, Zo, zooid, Zs, central disk (after *Bulman, Müller, Waterlot*).

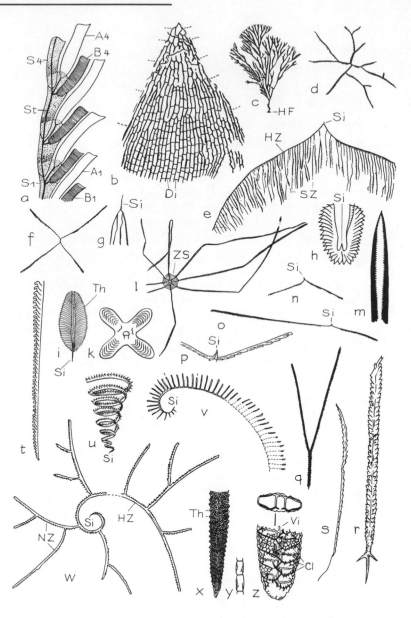

Several trends in the evolution of Graptoloidea are discernible, for example, the following:

1. Reduction in the number of the branches of a rhabdosome, from the original forty or more to a single branch with a single row of thecas.
2. Progressive straightening of the branches, whereby the nema became incorporated into the structure.
3. Increasingly complex thecas.
4. Reduction of the total number of thecas per rhabdosome (partly as a result of the first trend mentioned above).

In addition to these overall trends, the individual groups show their own specific developmental tendencies.

Graptolites proved to be reliable guide fossils for the Ordovician and Silurian periods, though of course they are primarily found in dark shales. They are rare or absent in sediments from shallow waters.

Graptolites preserved in calcareous rocks may be etched out of them with weak acids. Remains obtained in such fashion are eminently suited for detailed studies.

Conodonta (Conodontophorida, conodonts)

Cambrian–Upper Triassic (figure 64)

Conodonts are tiny (0.14 to about 4 mm long) tooth- or comb-shaped bodies of the carbonate apatite, frankolite. On a morphological basis,

FIGURE 63. Class Graptolithina. **a–d** Order Dendroidea. **e–z** Order Graptoloidea. **a** Schematic of a rhabdosome structure; **b** *Dictyonema flabelliforme,* 0.5x, Lower Ordovician; **c** *Dendrograptus,* 0.5x, Lower Ordovician; **d** *Anisograptus,* 0.5x, Lower Ordovician; **e** Pterograptus, 0.5x, Lower Ordovician; **f–h** *Tetragraptus,* 0.5x, Lower Ordovician; **i, k** *Phyllograptus,* Lower Ordovician, **i** Side view, 0.5x, **k** Ventral view, 2x; **l** *Dichograptus,* 0.2x, Lower Ordovician; **m–o** *Didymograptus,* 0.5x, Lower Ordovician; **p** *Leptograptus,* 1.5x, Ordovician; **q** *Dicranograptus,* 0.5x, Ordovician; **r** *Orthograptus,* 1x, Middle Ordovician; **s** *Dimorphograptus,* 1x, Lower Silurian; **t** *Monograptus priodon,* 1x, Lower Silurian; **u** *Monograptus turriculatus,* 1x, Lower Silurian; **v** *Rastrites,* 1x, Lower Silurian; **w** *Cyrtograptus,* 0.9x, Middle Silurian; **x–z** *Retiolites* (**x** 1x, **y, z** 6x). A, autotheca, B, bitheca, Cl, clathria, Di, dissepiments, HF, attachment area, HZ, main branch, NZ, side branch, S, stolotheca, Si, sicula, St, stolon, SZ, secondary branch, Th, theca, Vi, virgula, ZS, central disk (after *Bulman, Waterlot*).

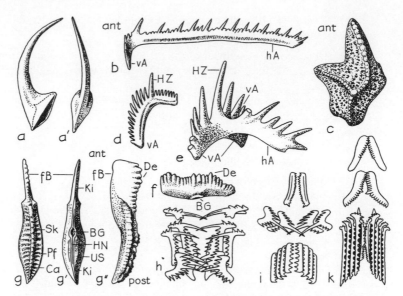

FIGURE 64. Conodontophorida. **a** *Drepano-dus,* 25x, Ordovician–Silurian; **b** *Hind-eodella,* 20x, Middle Silurian–Upper Triassic; **c** *Palmatolepis,* 12x, Upper Devonian; **d** *Prioniodina,* 12x, Sil-urian–Devonian, Triassic; **e** *Hibbar-della,* 15x, Middle Ordovician–Middle Triassic; **f** *Spatognathus,* 12x, Sil-urian–Middle Triassic; **g** *Polygnathus,* 15x, Devonian–Lower Carboniferous; **h–k** Restoration of conodont apparatus, Upper Carboniferous, **h** *Duboisella,* 10x, **i** *Illinella,* 10x, **k** *Scottognathus,* 8x. ant, anterior, BG, basal fossa, Ca, carina, De, denticle, fB, free blade, hA, posterior bar (process, limb), HN, attachment scar, HZ, main cusp, Ki, keel, Pf, platform, post, posterior, Sk, sculpture, US, crimp, vA, anterior bar (process, limb) (after *Hass*).

three types are usually distinguished: (1) individual, simple conodonts—distacodid forms (figure 64, a), occurring from Cambrian to Ordovician, (2) compound (or row) types (figure 64, d–f) with a bar- or blade-shaped base, a larger main cusp, and numerous small denticles; they occurred from Ordovician to Upper Triassic, and (3) platform conodonts (figure 64, c, g) whose denticles are carried on a broadened plate. The row of denticles may extend even past the platform. Platform conodonts were contemporary with the compound types (Ordovician to Upper Triassic).

Conodonts had a laminar structure and grew by addition of con-centric lamellae over a core. Well-preserved remains show that a conodont is in fact only the most resistant part of the so-called holoconodont, which consisted of a conodont sensu stricto and the basal body. During their growth the conodonts must have been enclosed in tissue, because there are indications that some broken parts had been regenerated.

A few rare finds demonstrate that several different individual conodonts together constituted a conodont apparatus (figure 64, h–k). Various conodonts appeared in such an assemblage, always in pairs (one or several), each pair consisting of two conodonts that were mirror images of each other. An apparatus encompassed between six and twelve conodont pairs, sometimes with additional, unpaired elements. As the knowledge of composition of the conodont apparatus increased, a new system was developed, that is, one based on the structure of the entire apparatus, in contrast to the earlier taxonomy, which was concerned solely with individual conodonts. Conodonts are found exclusively in marine sediments. The differences between types of conodonts found in different facies compel the conclusion that the conodont-bearing organisms were facies specific. In addition, different provinces can be distinguished. In contrast to the earlier assumption that the conodont-bearing organisms were pelagic, it is now generally believed that some have been very much substrate oriented. Theories about the nature and taxonomic status of the conodont-bearing animal were formerly quite divergent; recently, however, an impression of a roughly 40-mm-long, worm-shaped soft body was discovered in Lower Carboniferous deposits, with a conodont apparatus near one end. There is hardly any doubt that it is the remains of a conodont-bearing animal. The available material still does not permit classification of the organism in any of the known recent or fossil groups; therefore, the phylum Conodonta was established. It is probably related to chordates.

Conodonta, especially those of the platform type, emerged as good guide fossils, important for biostratigraphy of the Ordovician and up to the Upper Triassic period.

Trace Fossils

Precambrian–Recent (figure 65, table 13)

Trace fossils are preserved signs left in or on the substrate (carrier) by a living organism (generator). Intuitively, it would appear more suitable to classify these traces on the basis of the nature of their generators. However, because the generator is known only on the rarest of occasions, and, in addition, one organism may produce quite

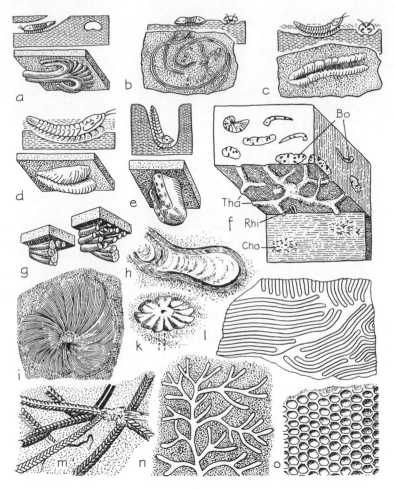

FIGURE 65. Trace fossils. **a–e** Behavior and trace types on the example of trilobites. **f–o** Various trace fossils. **a** Feeding passages, **b** Grazing traces, **c** Crawling traces, **d** Resting traces, **e** Dwelling traces. **f** Trace fossils in shale clays (below) and a limestone bank (above) of the Lower Lias, with perforated concretions and mussel shells; **g–k, n** Feeding structures; **l, o** Grazing traces, **m** Crawling traces; **g** *Phycodes,* 0.2x, Cambrian–Ordovician; **h** *Rhizo-* *corallium,* 0.3x, Cambrian–Tertiary; **i** *Zoophycos,* 0.2x, Devonian–Tertiary; **k** *Gyrophyllites = Palaeosemaeo-stoma,* 0.4x, Devonian–Tertiary; **l** *Helminthoida,* 0.7x, Cretaceous–Tertiary; **m** *Gyrochorte,* 0.9x, Middle Jurassic; **n** *Chondrites,* 1x, Cambrian–Tertiary; **o** *Palaeodictyon,* 0.3x, Ordovician–Tertiary. Bo, bore holes, Cho, *Chondrites,* Rhi, *Rhi-zocorallium,* Tha, *Thalassinoides* (after *Häntzschel, Seilacher, Schloz*).

TABLE 13.   Trace Fossils

Group	Subgroup
Immobility traces (Quietichnia)	Resting traces (Cubichnia) Dwelling structures (Domichnia)
Feeding traces (Cibichnia)	Feeding places and passages (Fodinichnia) Grazing traces (Pascichnia) Tooth or claw marks (Mordichnia) Hard substrate, or bore traces (Insolidichnia)
Locomotion traces (Movichnia)	Crawling traces (Repichnia) Running traces (Cursichnia) Walking traces (Gradichnia) Swimming traces (Natichnia) Flying traces (Volichichnia)
Traces of Metabolism and bioreactions	Excrement traces (Faecichnia) and traces interpreted as signs of illness, parasitism, or death throes

different traces, an ecological classification prevailed. Both trace fossils (= fossil traces) and the traces of recent animals can be categorized and subdivided according to the biological functions by which they were caused.

The various forms of the trace fossils are named parataxonomically. It is, therefore, easily possible that two different parataxa result from the activities of the same generator and, conversely, that different generators produced the same parataxa.

Trace fossils occur primarily in aquatic sediments. Many are very good facies indicators, having the added advantage of being always autochthonous. Their ecologic interpretation relies to a large degree on the results of actuopaleontological studies.

Several trace fossils are of certain importance as guide fossils as well. For example, various species of *Cruziana,* that is, the crawl tracks of trilobites, proved useful in the classification of sandy marine deposits from Old Paleozoic.

The Phylum Chordata

Next to the pulmonate mollusks and the arthropods, the chordates had solved the problem of terrestrial existence particularly well. Chordates are characterized by:

1. a notochord (chorda dorsalis), either throughout their life or only during embryonal stages;
2. a central nervous system (neural tube), situated above the notochord;
3. gill slits in the pharynx region, either persisting or only during embryonal stages;
4. a vascular system that is usually closed, with contractible sections.

Chordata encompass the following subphyla:

1. **Urochordata**, or **Tunicata**. Marine, in larval stages with a notochord and a dorsal neural tube. Adult sessile or pelagic; food intake is via a branchial sac.
2. **Cephalochordata** or **Acrania** (skull-less; *Branchiostoma* = *Amphioxus*); marine. The notochord reaches up to the front end of the body. Feeding via a branchial region of the pharynx. Epidermis single-layered.
3. **Vertebrata** or **Craniota**. Aquatic, amphibian, or terrestrial (figure 66). Body is divided into head, torso, and tail. There are two pairs of extremities, of which one or the other or even both may be absent. They are adapted to the mode of locomotion as fins, legs, or wings. Originally vertebrates have a persisting notochord and pharyngeal gill slits. The main sensory organs are concentrated at the front end of the dorsal neural tube. The epidermis has more than one layer.

Ontogeny and body organization of chordates point to their relationship with the phyla **Hemichordata** (Enteropneusta, Pterobranchia, † Graptolithina ?) and **Enchinodermata**. The phylogenetic relationships of the individual chordate subphyla are hypothetical, both with regard to each other and to other animal phyla. The closest relatives of chordates are the Cephalochordata. It is certain, however, that chordates developed from marine forms.

Subphylum Vertebrata

TERMINOLOGY

The following terms are used when referring to the location of various organs or parts of the body (figure 67):

Planes and Sections

Median (median sagittal) plane = plane of symmetry. Planes parallel to the median plane are sagittal planes (paramedian or longitudinal)

Transverse plane = plane perpendicular both to median and transverse planes

Orientation in Space

Cranial (anterior, oral) = at the head end
Caudal (posterior, aboral) = at the tail end
Medial = near the median plane (median = in the median plane)
Lateral = to the side, away from the median plane
Dorsal = back side
Ventral = belly side
Instead of the suffix -al, -ad may be used, meaning "in the direction of," for example, craniad = in the direction toward the head.
Central = toward the inside of the body
Peripheral = toward the outside, the body surface
Proximal = near the middle of the body
Distal = away from the middle of the body

Human anatomy uses somewhat different terminology in some cases.

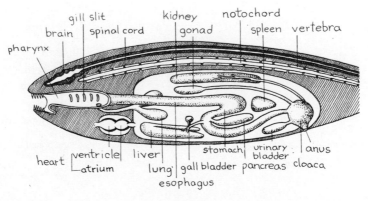

FIGURE 66. Diagram of the vertebrate body plan.

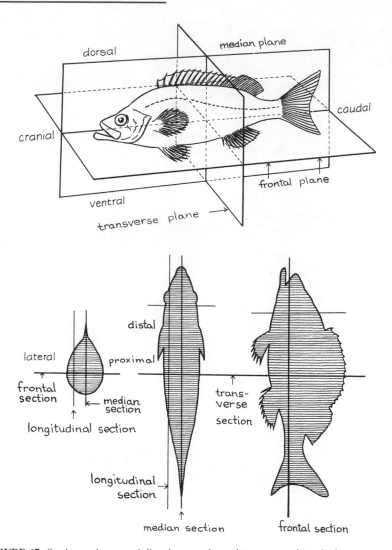

FIGURE 67. Sections, planes, and directions, as they relate to a vertebrate body.

Skeleton

The skeleton supports and protects the body, serves as a passive locomotion apparatus, provides anchor points for the muscles, and plays an important role in metabolism.

COMPOSITION

In general, only the hard parts of vertebrates are fossilizable, that is, bones and teeth and, rarely, cartilage. All these tissues develop from embryonal connective tissue, mesenchyme, of mostly mesodermal origin. Bones and cartilages of the visceral skeleton, certain parts of the skull base, and the tooth germs are derived from the neural ridge (figure 69).

CONNECTIVE TISSUE, CARTILAGE, AND BONE (figure 68)

Undifferentiated connective tissue consists of fibrocytes that produce an amorphous, gelatinous intercellular substance in which fibers may occur. Chondroblasts of the cartilage produce a transformable intercellular matrix that may calcify on the surface and thus become fossilized. During bone synthesis, osteoblasts secrete a basic matrix and are transformed into bone cells—osteocytes. The basic matrix becomes hardened through incorporation of hydroxyl apatite, $3(Ca_3P_4) \cdot Ca(OH)_2$. Unlike cartilage, bone is supplied with blood vessels and nerves, which accounts for its great biological plasticity. Two kinds of bone synthesis can be distinguished, based on the nature of the parent tissue. When bone is formed by replacement of cartilage, two parallel processes take place: during perichondral ossification, bone is added on the outside; during endochondral ossification, cartilage is destroyed and replaced by bone. Bones can also be formed from connective tissue, as cover, dermal, membrane, or secondary bones. Blood vessels play a significant role in ossification. Osteoblasts secrete layers of bone material in the form of hollow cylinders (Haversian system, osteon) around blood vessels. In many bones, compact bone (substantia compacta) is deposited toward the periphery and cancellous bone (substantia spongiosa) toward the inside, so that both opposing requirements are met: great strength and light weight. The cancellous structure complies with the mechanical requirements; the thickness ratios of bones follow rules of simple mechanics.

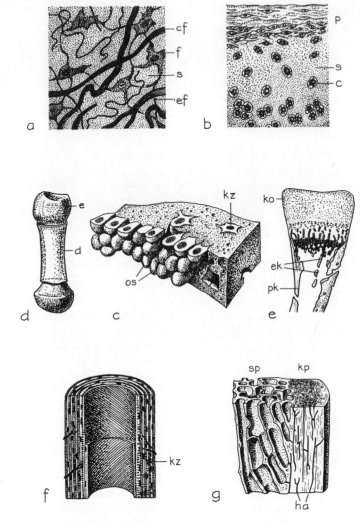

FIGURE 68. Structural elements of the vertebrate body. **a** Loose connective tissue; cf, collagen fibers, ef, elastic fibers, f, fibrocyte, s, intercellular substance. **b** Hyaline cartilage; c, chondroblast, p, perichondrium, s, intercellular substance. **c** Bone formation; kz, bone cell (osteocyte), os, osteoblast. **d** Diagram of a long bone; d, diaphysis, e, epiphysis. **e** Longitudinal section through epiphysis and diaphysis of a long bone; ek, enchondral ossification, ko, cartilage, pk, perichondral ossification. **f** Diagram of the Haversian system (osteon); kz, osteocyte. **g** Piece of a long bone; ha, Haversian canals, kp, compacta, sp, spongiosa (after *Braus, Grassé*).

DENTINE, ENAMEL, AND CEMENT (figure 69)

Like bone tissue, dentine, enamel, and cement (cementum) consist of a varying proportion of inorganic material (primarily hydroxylapatite) and collagen fibers. These hard substances were first studied in mammals. In a tooth germ that is derived from the neural crest, odontoblasts secrete dentine (substantia eburnea) around a pulp cavity. From the epidermis a dental ridge grows inward, producing enamel organs over the tooth germs. An epithelium of adamantoblasts (enamel builders) secretes the enamel inward. Enamel (substantia adamantina) is a particularly hard material made of crystalline apatite with prismatic structure. It occurs only in mammals; the term (enamel) cannot be applied in its specific meaning to any of the lower vertebrates. The enamellike, hypermineralized outer tooth layer of many fishes is of dentine origin. It is then designated as enameloid (vitrodentine, durodentine, hyalodentine). In addition to the usual orthodentine, there is also trabecular dentine, which forms the network inside the pulp cavity, and vasodentine, which contains blood vessels. Cement (or cementum) is a bony tissue that contains very few cells; in mammal teeth it may cover either the root area or the entire tooth.

AXIAL SKELETON

The axial skeleton of vertebrates consists of the anteriorly located skull and attached spinal column.

SKULL (figure 70)

Neuro- and Splanchnocranium

A head capsule, the neurocranium, at the frontal pole of the body, contains brain and main sensory organs. The adjoining skeletal apparatus arises from the visceral arches. This jaw and gill skeletal system or apparatus is termed splanchnocranium or viscerocranium. Neuro- and splanchnocranium are both cartilaginously founded in the embryo as primordial- or chondrocranium that later partly ossifies. In most cases the chondrocranium becomes enveloped in a dermal bone armor known as dermatocranium. All skull components have a long phylogenetic history. The morphology of the skull can thus be properly understood only through consideration of both its ontogenetic and phylogenetic development. The close relationship of the skull and

154

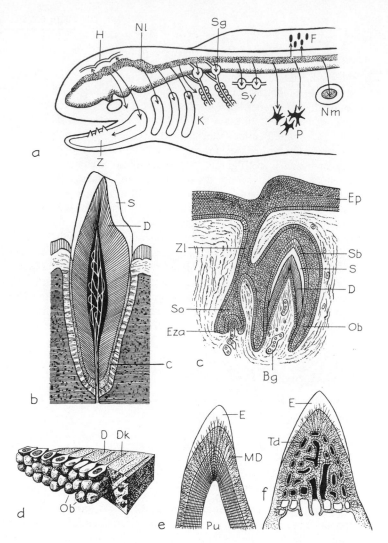

FIGURE 69. a Diagram of derivatives of the neural crest of an amphibian embryo; F, finfold epithelium, H, pericranium, K, cartilage of the splanchnocranium, Nl, neural crest or ridge, Nm, medulla of the suprarenal gland, P, pigment cells, Sg, spinal ganglion, Sy, sympathicus ganglion, Z, tooth germs. **b** Section through a mammal tooth; C, cementum, D, dentine, S, enamel. **c** Section through dental ridge of a mammal embryo, showing a milk tooth and a replacement tooth germ; Bg, blood vessels, D, dentine, Ep, epithelium, Eza, replacement tooth germ, Ob, odontoblasts, S, enamel, Sb, adamantoblasts (enamel producers), So, enamel organ, Zl, tooth ridge. **d** Secretion of dentine by the odontoblasts; D, dentine, Dk, dentine canaliculi, Ob, odontoblasts. **e, f** Longitudinal sections through fish teeth; E, enameloid (durodentine), MD, mantle dentine, Pu, pulp cavity, Td, trabecular dentine (after *Hadorn, Hildebrand, Kuhn, Rietschel*).

corresponding soft parts allows inferences about brain structure, the course of nerves and blood vessels, and head musculature, based on the study of hard parts.

The structural material of neurocranium is of various origins (mesodermal and mesectodermal). Splanchnocranium has a special position in cyclostomes, but in all gnathostomes it is built on a uniform plan (figure 70, a). In the front is a modified visceral arch, the mandibular arch. It consists of dorsal palatoquadratum and the Meckel's cartilage (mandibular); it serves as the food intake. The next arch, the hyoid, consists of the dorsal hyomandibular and ventral hyoid. Adjacent visceral arches are comprised of four components on each side and an unpaired connecting piece. Two arches border one slit. Between the mandibular and hyoid arches is the spiracle (spiraculum), a modified gill slit. The middle ear of tetrapods is homologous with the spiracle of fishes. Of particular importance is the type of attachment of the mandibular to the neurocranium (jaw suspension): amphistylic, hyostylic, or autostylic (figure 70, b–d). With the exception of amphibian larvae, the mandibular arches of tetrapods are reduced; their remains are associated with the tongue and the trachea.

Cranial Skeleton and Brain

The central nervous system of vertebrates is divided into the following regions (figure 70, e–f):

1. prosencephalon (forebrain)
   a) telencephalon
   b) diencephalon
2. rhombencephalon (hind brain)
   a) mesencephalon
   b) metencephalon
   c) myelencephalon
3. medulla oblongata (spinal cord)

Prosencephalon: The telencephalon, paired eversions of the prosencephalon, originally performed a purely olfactory function. During phylogeny the centers of association became progressively differentiated until, in the mammals, the prosencephalon became the location of the greatest mental capabilities. Parts of the roof of the diencephalon were differentiated into light-sensing organs (pineal and parietal eyes). The hypophysis is located underneath.
Rhombencephalon: Mesencephalon is the dominant association center of lower vertebrates; in the higher vertebrates it becomes an important "switchyard" of neural impulses. The most important dorsal

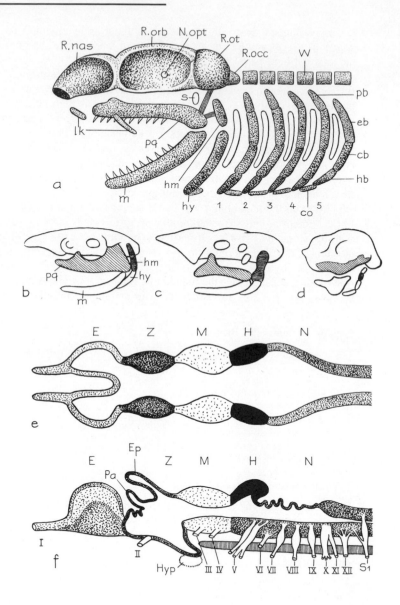

structure of the metencephalon is the cerebellum, the center for coordination of locomotion and muscle tone. Myelencephalon is similar in structure to the spinal cord. It contains the centers of vegetative functions (breathing, metabolism, vascular innervation, sleep).

Cranial nerves: As in human anatomy, the cranial nerves are numbered from front to rear, regardless of function. Amniota possess the following twelve pairs:

  I. N. olfactorius, olfactory nerve; sensory
  II. N. opticus, optic nerve; sensory
  III. N. oculomotorius, ocular muscle nerve
  IV. N. trochlearis, ocular muscle nerve
  V. N. trigeminus, trigeminal nerve, branchial; jaw musculature
  VI. N. abducens, ocular muscle nerve
  VII. N. facialis, facial, branchial nerve; organs of taste, musculature of the hyoid arch, mimic muscles in humans
  VIII. N. statoacusticus (N. vestibulocochlearis), statoacustic, sensory, inner ear
  IX. N. glossopharyngeus, glossopharyngeal, branchial nerve; muscles of the pharynx, salivary glands, mouth and tongue mucous membranes
  X. N. vagus, branchial nerve; in fishes, innerving all branchial arches; one branch (Ramus visceralis) continues to the intestines
  XI. N. accessorius Willisii, accessory, branchial nerve; branchial musculature
  XII. N. hypoglossus, hypoglossal; three spinal nerves without dorsal roots, tongue muscles

FIGURE 70. a Chondrocranium of a shark. Areas of the neurocranium: R.nas, regio nasalis, R.orb, regio orbitalis, R.ot, regio otica, R.occ, regio occipitalis, N.opt, nervus opticus. Splanchnocranium: lk, lip cartilage, pg, palatoquadrate, m, Meckel's cartilage (mandibular), hm, hyomandibular, hy, hyoid, s, spiracle. 1–5 branchial arches; pb, pharyngobranchial, eb, epibranchial, cb, ceratobranchial, hb, hypobranchial, co, basibranchial (copula), W, vertebra. **b–d** Main types of jaw suspension in fishes: **b** amphistyly, **c** hyostyly, **d** autostyly (holostyly); hm, hyomandibular, hy, hyal, m, Meckel's cartilage, pq, palatoquadrate. **e, f** Diagram of vertebrate brain, **e** Frontal section, **f** Sagittal section showing the cranial nerves (I–XII); S1, 1st spinal nerve, Prosencephalon: E, telencephalon, Z, diencephalon. Rhombencephalon: M, mesencephalon, H, metencephalon, N, myelencephalon; Ep, epiphysis (pineal organ), Pa, parapineal organ (parietal organ, parietal eye), Hyp, hypophysis (after *Kühn, Starck*).

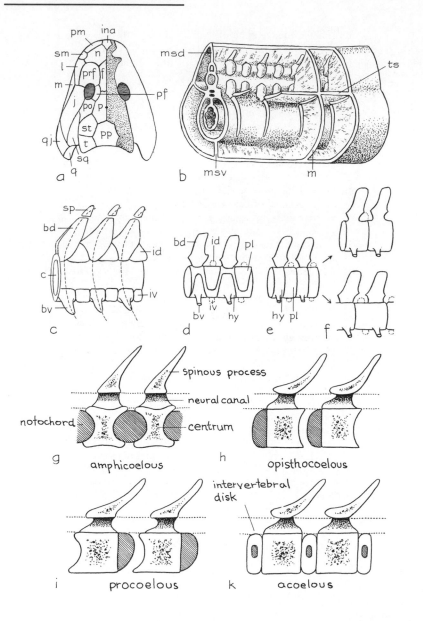

From the structure of the twelfth cranial nerve it can be inferred that the rear part of the skull of higher vertebrates (Amniota) is equivalent to three body segments fused to the occiput. Living amphibians lack the hypoglossal, which, however, was present in many fossil Amphibia (Labyrinthodontia). The nature of the posterior end of the fish skull has in fact not been properly determined.

Dermatocranium (figure 71, a)

Fossil Agnatha and Placodermi show the specific number and arrangement of the components of their dermal head armor. In bony fishes and in tetrapods the dermal bones form a secondary roof over the neurocranium, with openings for sensory organs. The dermal bones exhibit a more or less regular bilaterally symmetrical pattern. One fundamental plan can be discerned and followed from the Crossopterygii up to the mammals. It is difficult to relate this pattern to the Actinopterygii, and the situation of Dipnoi falls entirely outside the scheme. The lateral line system, where the relationships are somewhat more constant, is important for the identification of individual skull parts and for the establishment of homologies in fishes. However, the number and arrangement of the dermal components can be quite variable. In tetrapods the pattern is more constant.

SPINAL COLUMN (figures 71, 72)

The spinal (or vertebral) column consists of notochord and perichordal skeleton. The notochord, a flexible rod, is phylogenetically older. In the vertebrates it reaches from the rear end of the body up

FIGURE 71. a Skull of a primitive tetrapod (*Ichthyostega*), left, dermatocranium, right (dotted), endocranium; f, frontal, internasal, j, jugal, l, lacrimal, m, maxilla, n, nasal, p, parietal, pf, postfrontal, pm, premaxilla, po, postorbital, pp, postparietal, prf, prefontal, q, quadrate, qj, quadratojugal, sm, septomaxilla, sq, squamosal, st, supratemporal, t, tabular. **b** Schematic of the vertebrate connective tissue system; m, transversal septum, msd, median dorsal septum, msv, median ventral septum, ts, horizontal septum. **c** Aspondylic stage of the vertebral column (*Acipenser*); bd, basidorsal, bv, basiventral, c, notochord, id, interdorsal, iv, interventral, sp, spinal. **d–f** Types of vertebrae: **d** hemispondylic stage, **e** diplospondylic stage, **f** monospondylic stage; hy, hypocentrum = intercentrum, pl, pleurocentrum. **g–k** Articular surfaces between adjacent vertebrae (after *Goodrich, Jarvik, Remane*).

to the hypophysis, under the skull base. In the cyclostomes and in many fishes it persists throughout their life. However, in many fishes it is already partially supplanted by the vertebral body, and in tetrapods it has lost its importance as a support organ. Its remnant is the nucleus pulposus, a gelatinous substance inside the intervertebral disks.

The notochord and spinal cord as well as the blood vessel located below the notochord are, in *Amphioxus*, enveloped in a perichordal skeleton of connective tissue; in vertebrates this skeleton

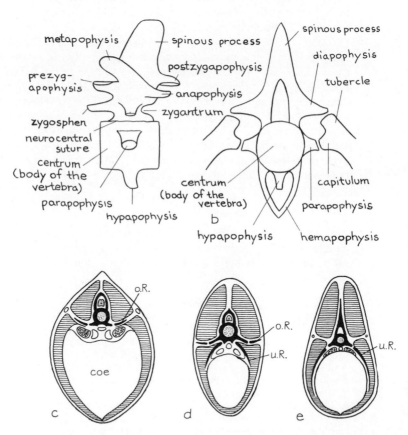

FIGURE 72. a, b Vertebral processes. **c–e** Schematic cross sections of bodies of fishes. Position of the ribs: **c** Selachii, **d** Crossopterygii, **e** Teleostei; coe coelom, o.R., dorsal rib, u.R., ventral (pleural) rib (after *Bütschli*).

consists of cartilaginous or bony components. Phylogenetically, there are various distinct stages.

Aspondylic stage: Arch components (arcualia) surround the neural tube, adjacent to the notochord, and two dorsal and two ventral arcualia can be distinguished in each segment. A vertebral body as such (centrum) is absent (figure 71, c):

1. dorsal anterior arcuale = interdorsal (dorsal intercalar piece),
2. dorsal posterior arcuale = basidorsal (neural arch),
3. ventral anterior arcuale = interventral (ventral intercalar piece),
4. ventral posterior arcuale = basiventral (hemal arch).

Development of the vertebral body: The body develops partly from the arch components (arcocentrally) and partly independently from the perichondral connective tissue (autocentrally). In Selachii, the fibrous sheath of the notochord is involved as well (chordacentral development).

Hemispondylic development: A ventral bony half-ring (intercentrum = hypocentrum) and a dorsal bony half-ring (pleurocentrum) are located next to the notochord in each segment. The neural arch (basidorsal) is situated over the intercentrum (figure 71, d).

Holospondylic development: The intercentrum and pleurocentrum expand, each developing into one centrum that encases the notochord, so that two vertebral bodies are formed per segment (primary diplospondyly, figure 71, e). However, most species have only one vertebral centrum per segment (are monospondylic). This can be achieved in three ways (figure 71, f):

1. by fusion of intercentrum and pleurocentrum (concrescent centrum),
2. by reduction of the pleurocentrum and expansion of the intercentrum (expansion centrum),
3. by reduction of the intercentrum and expansion of the pleurocentrum (expansion centrum).

During the ontogeny, vertebrae assume intersegmental locations, functionally advantageous for muscle attachment.

Shapes of the Vertebral Body

The vertebral body (centrum) may displace the notochord either partially or entirely. The adjacent articular surfaces of two bodies may have different shapes (figure 71, g, k). There are five main types:

1. Amphicoelous: the centrum is concave at both ends; remnants of the notochord, sometimes connected together by a canal, are usually preserved in such primitive vertebrae.
2. Procoelous: anterior articular face is concave, posterior convex.

3. Opisthocoelous: anterior face is convex, posterior concave.
4. Platycoelous (acoelous): both terminal faces are flat.
5. Heterocoelous: both faces are saddle shaped.

Vertebral Processes are primary objects of study of functional anatomy (figure 72, a, b).

1. Zygapophyses: processes that serve to join together two subsequent vertebrae:

   a) Pre- and postzygapophyses of the neural arch: the articular surfaces of the prezygapophyses generally face upward and link with the downward-pointing surfaces of postzygapophyses. They are fully developed only in tetrapods.
   b) Zygosphen-zygantrum: unpaired pin-joint between the neural arches. The pin (zygosphen) rises from the anterior.
   c) Hyposphen-hypantrum: in shape, much like the zygosphen-zygantrum, except that the pin (hyposphen) projects posteriorly from the neural arch.

   Zygosphen-zygantrum and hyposphen-hypantrum joints are found only in certain amphibians and reptiles. They prevent torsion of the spinal column.

2. Processes serving the attachment of ribs:

   a) Basapophyses: for the attachment of ventral ribs. In the caudal region the basapophyses merge with the hemal arch (hemapophyses, chevrons).
   b) Epapophyses: in fishes, lateral processes for the rib attachment.
   c) Parapophyses: processes to link with the capitula of two-headed (dichocephalic, bicipital) ribs. Only in tetrapods.
   d) Diapophyses: to link with the tubercle of bicipital ribs.
   e) Synapophysis: a fusion product of para- and diapophysis.
   f) Pleurapophysis: a rounded-off processus for attachment of a rudimentary rib.

3. Spinal processes: dorsal or ventral median processes extending from the centrum.

   a) Dorsal process (processus spinosus): projects from the neural arch into the median dorsal longitudinal septum. In fishes it may be formed of separate components (spinalia). In tetrapods it may interact with the dermal skeleton.
   b) Hypapophyses: ventromedian processes of the centrum, serving as a means of muscle attachment.

Other processes, also generally related to muscle attachment, may occur in various locations on the centrum.

Ribs

Ribs are segmental, rod-shaped skeletal components that are embedded in the connective tissues of the dividing walls (septa) or in the lateral muscles of the torso. They attach proximally to the vertebral column; the distal ends are either free or joined with the sternal structures. Originally, ribs occurred from the neck down to the caudal region. In mammals and birds, well-developed ribs are limited to the thoracic region. Dorsal and ventral (pleural) ribs may be distinguished (figure 72, c–e), but usually only one set is developed. In contrast to fishes, tetrapod ribs are normally forked at the proximal end. Dorsally the costal tubercle links with the diapophysis of the vertebra; ventrally the capitulum links with the parapophysis. In tetrapods, one-headed ribs can derive in various ways from the bicipital ones. Abdominal ribs (gastralia) of the lower tetrapods are of dermal origin.

VERTEBRATE TAXONOMY

Credit for the introduction of the term Vertebrata belongs to J. B. LAMARCK (1794). He used it to replace ARISTOTLE's designation of "animals with blood" and LINNAEUS's term "red-blooded animals." Vertebrates are the most advanced subphylum of the Chordata. The subphylum can be divided into two superclasses, Agnatha (without jaws) and Gnathostomata (jaw mouth), with several classes. Since their specialization followed a different direction, Gnathostomata cannot be derived directly from any of the known Agnatha. Agnatha comprise four classes, and Gnathostomata eight. There are two principal groups of the Gnathostomata: Pisces (fishes) and Tetrapoda (four-footed). The three highest classes (Reptilia, Aves, Mammalia) can, as Amniota, be distinguished from the Anamnia. They possess special embryonal organs (amnion, chorion, allantois) and do not lay eggs in water. Reptilia and Aves are frequently grouped together as Sauropsida. We use the classification outlined in table 14, and, in a simplified form, in table 15.

1. Superclass Agnatha (jawless)

The superclass Agnatha is represented today by the cyclostomes, which include the parasitic lampreys and the scavenging Myxinoidea (figure 73, a, b). They have fossil relatives, many of which possessed a bony exoskeleton. During the older Paleozoic (Cambrian–Devonian), their distribution was cosmopolitan. For a bet-

ter understanding of the organization of Agnatha, the living representatives are characterized first:

1. Jaws originally absent. Body eel shaped, without paired fins.
   Skin naked, with many glands.

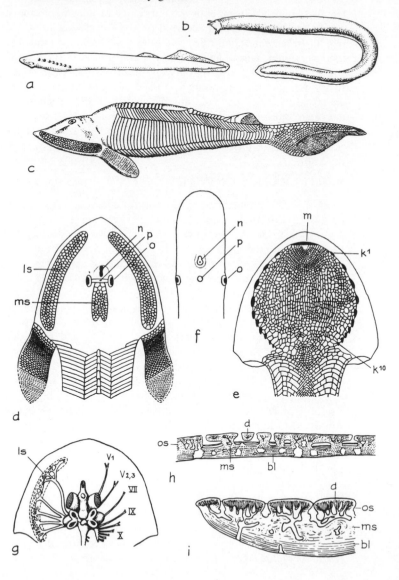

TABLE 14.   Classification of Vertebrates

	Classes
Superclass Agnatha	Myxinomorpha Recent
	Cephalaspidomorpha Silurian–Recent
	Pteraspidomorpha Cambrian–Recent
	Thelodonti Silurian–Devonian
Superclass Gnathostomata	
Pisces (fishes)	Placodermi Silurian–Carboniferous
	Chondrichthyes (cartilaginous fishes) Silurian–Recent
	Acanthodii Silurian–Permian
	Osteichthyes (higher bony fishes) Silurian–Recent
Tetrapoda (tetrapods)	Amphibia (amphibians) Devonian–Recent
Amniota	Reptilia (reptiles) Carboniferous–Recent
	Aves (birds) Jurassic–Recent
	Mammalia (mammals) Triassic–Recent

FIGURE 73. Agnatha. **a** *Petromyzon marinus*, sea lamprey. **b** *Myxine glutinosa*, hag fish. **c–e** *Hemicyclaspis* (Lower Devonian, Osteostraci), **c** Restoration, approx. 20 cm long, **d** Dorsal view of the head shield, **e** Ventral view of the head shield; **f** *Petromyzon* (Recent) dorsal view of the head; K1–K10, outer openings of the gill slits, ls, lateral sensory field, ms, median sensory field, m, mouth, n, nasohypophyseal opening, o, orbita, p, pineal opening. **g** *Kiaeraspis* (Lower Devonian, Osteostraci). Cast of the skull cavities; V_1, nervus profundus, $V_{2,3}$, nervus trigeminus, VII, nervus facialis, IX, nervus glossopharyngeus, X, nervus vagus, ls, lateral sensory field. **h** *Tremataspis* (Upper Silurian, Osteostraci). Cross section of the exoskeleton. **i** *Corvaspis* (Upper Silurian, Heterostraci), cross section of the exoskeleton; bl, basal layer, d, dentine, ms, middle layer, os, upper layer (after *Denison, Grassé, Gross, Norman, Stensiö*).

TABLE 15.   Simple Classification of Vertebrates

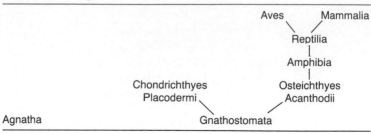

2. Endoskeleton of hyaline cartilage and cartilaginous tissue. Neural endoskeleton. Branchial cavity with five to twelve pairs of gills in loose pockets. Unpaired nasohypophyseal duct. Mouth suctorial, with horny epidermal teeth.
3. Notochord persisting.
4. Differentiated brain with eight or ten pairs of cranial nerves. Dorsal and ventral roots of the spinal nerves are separated (Petromyzontoidea). Eyes either well developed (Petromyzontoidea) or reduced (Myxinoidea). Labyrinth with only one (Myxinoidea) or with two semicircular canals (Petromyzontoidea). Pineal and parietal organs present.
5. Heart has atrium and ventricle, handles only venous blood.
6. Variable body temperature (poikilothermy).
7. Fertilization external; development is either direct (Myxinoidea), or there is a prolonged larval stage (*Ammocoetes* of the Petromyzontoidea).

Substantial differences between Myxinoidea and Petromyzontoidea suggest that a split between the two groups occurred a very long time ago. Marine Myxinoidea are the more primitive forms.

Fossil Agnatha ("Ostracoderms") differed from the recent representatives by having a bony exoskeleton. It is assumed that the cyclostomes had lost the ability to build bone. A proof of kinship was obtained by meticulous analysis of the head region of Osteostraci (figure 73, d–f). Currently the taxonomy of Agnatha is again unsettled. We distinguish the following classes:

Superclass Agnatha

1. Class Myxinomorpha (up to now only recent forms are known)
2. Class Cephalaspidomorpha
   1. Subclass Osteostraci (Middle Silurian–Upper Devonian)
   2. Subclass Anaspida (Lower Silurian–Upper Devonian)

3. Subclass Petromyzonta (Carboniferous–Recent)
3. Class Pteraspidomorpha
   1. Subclass Heterostraci (Upper Cambrian–Middle Devonian)
   2. Subclass Galeaspida (? Lower to Middle Devonian)
4. Class Thelodonti (Lower Silurian–Upper Devonian)

Myxinomorpha are represented only by *Myxine* and its living relatives.

In **Osteostraci** (figure 73, c), the head and the anterior parts of the body are enclosed in a multilayered bony armor. The sensory fields, one paired and one single, are related to the lateral line and are located dorsally on the head shield. Structures similar to pectoral fins may occur. The tail is heterocercal. Anaspida (figure 74, a, b) are small, active forms with fusiform bodies and hypocercal tails. The isolated fossil Petromyzonta known to date are from the Carboniferous (*Mayomyzon, Hardistiella*); they are similar to modern lampreys.

The bony armor of **Heterostraci** consists of a variable number of shields. There is a common opening over the gills. Eyes are small; the tail is diphycercal. Fragments of Heterostraci exoskeletons have been documented since Upper Cambrian. **Galeaspida** possess massive skeletal shields. To date, they are known only from the Devonian of China.

Thelodonti (figure 74, f) have bodies covered with small scales or toothlets. They have a dorsal fin, an anal fin, and lateral "fin flaps." The tail is hypocercal. Whether Thelodonti represent a natural group is questionable. They may have given rise to some of the Agnatha, and maybe Gnathostomata (??).

2. Superclass Gnathostomata

Most fossil and recent vertebrates belong to the gnathostomes. They are characterized by the presence of jaws, which are modified branchial arches. All Gnathostomata have labyrinths with three semicircular canals.

Pisces (fishes) (figure 75)

Initially, fishes had two pairs of fins. The origin of fins is unknown, and none of the extant theories is fully satisfactory. The theory of

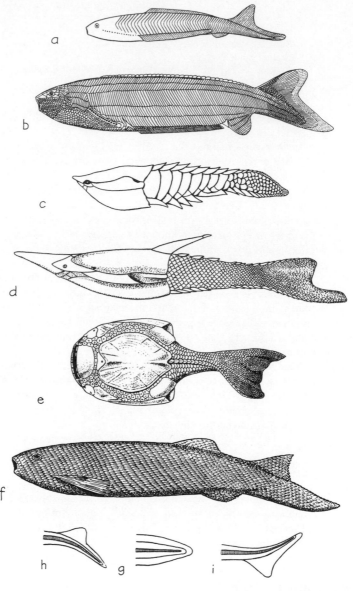

FIGURE 74. a, b Anaspida: **a** *Jamoytius* (Upper Silurian), ≈21 cm, **b** *Pharyngolepis* (Upper Silurian), ≈22 cm. **c–f** Heterostraci: **c** *Anglaspis* (Lower Devonian), ≈6 cm, **d** *Pteraspis* (Lower Devonian) ≈24 cm, **e** *Drepanaspis* (Lower Devonian), ≈24 cm, **f** Thelodonti: *Phlebolepis* (Upper Silurian), ≈6 cm. **g–i** Types of tail fin: **g** protocercal, **h** hypocercal, **i** heterocercal (after *Gross, Kiaer, Ritchie, Starck, White*).

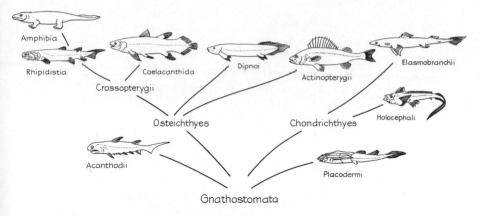

Amphibia

Rhipidistia   Coelacanthida   Dipnoi   Elasmobranchii

Crossopterygii   Actinopterygii

Osteichthyes   Chondrichthyes   Holocephali

Acanthodii

Placodermi

Gnathostomata

FIGURE 75. Phylogenetic tree of the lower Gnathostomata.

lateral folds, which has received the most support, assumes that the paired fins are derived from bilaterally developed skin folds.

Extant fishes form two large natural groups: Chondrichthyes (cartilaginous fishes: sharks and chimaeras) and Osteichthyes (bony fishes). Two additional groups lived during the Paleozoic: Placodermi and Acanthodii. Certain placoderms (Ptyctodontida) resembled modern chimaeras; acanthodes were probably related to the Osteichthyes.

1. Class Placodermi (figure 76)

Placoderms of the Devonian are an early experiment of the Gnathostomata. They are quite variable and are generally small or of moderate size. They may be characterized as follows:

1. Head with a head shield that is hinged together with the thoracic shield. Posteriorly the body tapers off, ending usually with a heterocercal caudal fin.
2. Each order exhibits a characteristic pattern of the dermal bone elements of the head and body shield, not discernibly homologous with those of higher animals. Jaw suspension is auto- or hyostylic. The masticatory organ consists of modified dermal bone plates. The lateral line system is well developed.
3. The persisting parts of the notochord are not connected. Vertebrae consist only of neural and hemal arches. The foremost neural arches are fused into a synarcuale.

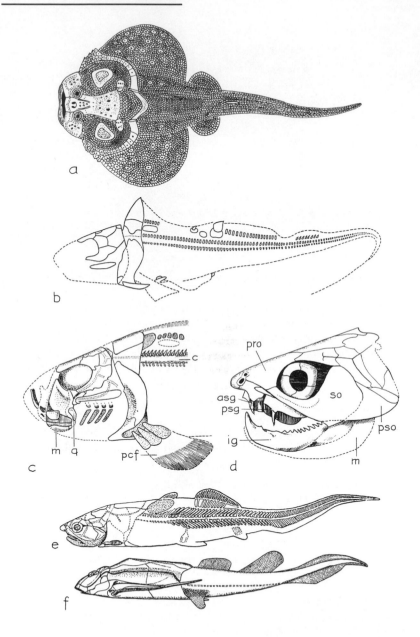

4. Pectoral, pelvic, and one or two dorsal fins are developed. The animals are mostly bottom dwellers, in freshwater or marine environments.

Classification of Placodermi is difficult because of their variable, often remarkable adaptations. Only four orders can be isolated. Order **Rhenanida** encompasses dorsoventrally flattened, raylike fishes with large pectorals (*Gemuendina*). Representatives of the order **Ptyctodontidae** remotely resemble the recent chimaeras. They are small, predominantly marine fishes; males have pterygopods similar to those of cartilaginous fishes. Best known are *Rhamphodopsis* and *Ctenurella*. Order **Arthrodira,** with both small and very large representatives, accounts for more than 70% of all known placoderms. Very large forms became known from the Devonian; for example, the head shield of *Titanichthys* is up to 60 cm long. Order **Antiarchi** contains highly specialized, small to middle-sized fishes with a cosmopolitan distribution. For genus *Bothriolepis,* it was possible to restore some of the anatomy of the soft parts.

2. Class Chondrichthyes (cartilaginous fishes)

Chondrichthyes are mostly marine, predatory fishes with a cartilaginous endoskeleton. Living representatives can be separated into two quite different groups:

1. Subclass Elasmobranchii
   Order Selachii (Pleurotremata), sharks
   Order Batoidea (Hypotremata), rays and skates
2. Subclass Holocephali
   Order Chimaerida, chimaeras

The only remains of fossil Chondrichthyes are teeth, fin and head spines, and scales. Recent Chondrichthyes are characterized as follows (figure 77):

FIGURE 76. Placodermi. **a** *Gemuendina* (Lower Devonian, Rhenanida), ≈25 cm. **b** *Rhamphodopsis* (Middle Devonian, Ptyctodontida), ≈12 cm. **c** *Ctenurella* (Upper Devonian, Ptyctodontida), restoration of the head, ≈3 cm; c, notochord, m, mandibular, pcf, pectoral fin, q, quadrate. **d** *Hadrosteus* (Lower Devonian, Arthrodira), skull, ≈16 cm; asg, anterior supragnathal, ig, infragnathal, m, mandibular, pro, preorbital, psg, posterior supragnathal, pso, postsuborbital, so, suborbital. **e** *Coccosteus* (Middle Devonian, Arthrodira), ≈22 cm. **f** *Bothriolepis* (Upper Devonian, Antiarchi), ≈20 cm (after *Denison, Gross, Miles, Ørvig, Stensiö*).

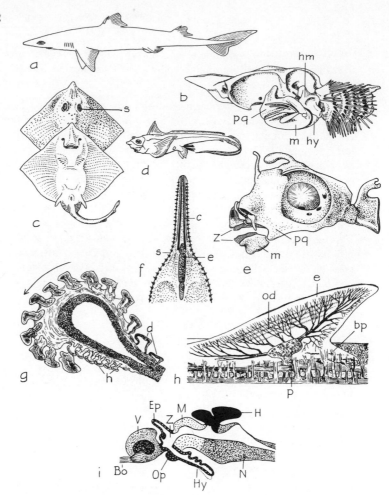

FIGURE 77. Body organization of Chondrichthyes. **a** *Squalus,* spiny dogfish. **b** *Squalus,* skull; hm, hyomandibular, hy, hyoid, m, Meckel's cartilage (mandibular), pq, palatoquadrate, **c** *Raia,* ray (Recent), s, spiracle behind the eye. **d** *Chimaera* (Recent), male with pterygopodia. **e** *Chimaera,* skull, branchial arch removed; m, Meckel's cartilage (mandibular), pq, palatoquadrate, Z, dental plates. **f** Cross section of the dorsal fin of a shark; c, ceratotrichia, (horny fin rays), e, skeletal support of fin, s, scale. **g** Section through lower jaw of *Mustelus,* dog shark (Recent), revolving dentition; the arrow shows direction of the tooth replacement; d, replacement teeth, h, dermal teeth. **h** Longitudinal section of a placoid scale of *Scymnus* (Recent); bp, basal plate, e, enameloid, od, orthodentine, p, pulp. **i** Shark brain; Bo, bulbus olfactorius, Ep, epiphysis, H, metencephalon, Hy, hypophysis, M, mesencephalon, N, myelencephalon, Op, entrance of the optic nerve, V, telencephalon, Z, diencephalon (after *Goodrich, Hertwig, Kühn, Marinelli, Norman, Starck*).

1. Skin is generally protected by an exoskeleton of small placoid scales that consist of dentin with a bony base.
2. The cartilaginous endoskeleton may ossify on the surface. Jaw suspension amphi-, hypo-, or autostylic (Holocephali). Notochord persisting. In males the distal parts of the pelvic fins are modified into copulatory organs (pterygopodia). Caudal fin heterocercal or diphycercal (Holocephali).
3. Heart with atrium and ventricle, venous blood driven into five to seven pairs of gills (Holocephali, four pairs). Gill slits are either open directly to the outside or covered with an operculum (Holocephali).
4. No lungs and no swimbladder.
5. Brain hemispheres, olfactory lobes (macrosmatic), and lateral line well-developed.
6. Body temperature variable (poikilothermy).
7. Internal fertilization, oviparity or ovoviviparity. Eggs large, rich supply of yolk. Development direct, without metamorphosis.

Distribution: Upper Silurian–Recent

Bone tissue occurs in the Chondrichthyes only in bases of the placoid scales. Its absence is interpreted either as primary or as retrogressive. The period of greatest prosperity for Chondrichthyes was the Carboniferous. Some finds from recent years indicate that the division between Elasmobranchii and Holocephali may not have been as sharp as shown by the recent representatives. This applies mainly to the ones closest to the Holocephali. On these grounds R. ZANGERL has proposed a new subclass, Subterbranchialia, in place of the Holocephali. However, it is questionable whether this subclass represents a natural unit.

1. Subclass Elasmobranchii (figure 78, a–k)

Elasmobranchii include excellent swimmers as well as bottom-dwelling forms (Batoidea). Their main branch begins in the Upper Devonian with primitive **Ctenacanthoidea** that had split from **Hybodontoidea** during the Mesozoic. They have mostly sharp, pointed teeth, but wide, stubby ones may occur as well. Ensuing modern sharks (**Euselachii**) have subterminal mouths, vertebrae with centrums, and fins with narrow bases; the oldest known representative is *Palaeospinax* (Lower Jurassic). Among Euselachii are distinguished **Galeoidea** and **Squaloidea**. The majority belong to the Galeoidea, with five pairs of gill slits. Fossil remains include mainly teeth.

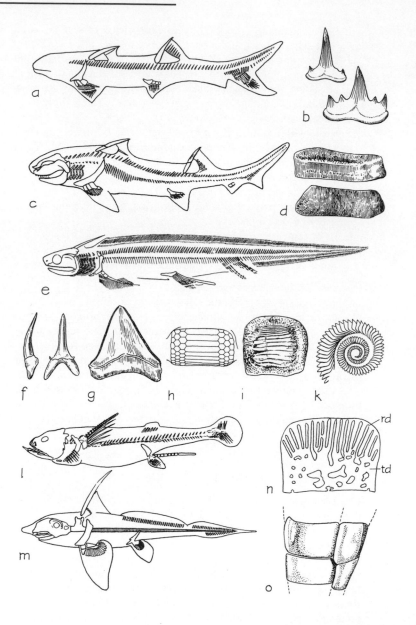

Squaloidea show a tendency toward a benthic way of life. An example is the angel shark (*Squatina*), with its very much enlarged pectoral fins.

Members of the order **Batoidea** (rays), mostly benthic, have been known since the Jurassic. Their dentition is made up of flat teeth that serve to break open hard-shelled prey.

A number of extinct orders must be mentioned, in addition to Selachii and Batoidea. Representatives of the ancient order **Cladoselachida** (Devonian–Carboniferous) had teeth with a central point flanked by secondary points. Members of the aberrant order **Xenacanthida** (Devonian–Triassic) possessed head spines and pectoral fins with a biserial archipterigium. Besides these there were further orders (**Symmoriida**, Carboniferous; **Eugeneodontida**, Carboniferous–Triassic; **Orodontida**, Carboniferous; **Petalodontida**, Carboniferous–Permian).

2. Subclass Subterbranchialia (figure 78, 1–o)

In contrast to the Elasmobranchii, the pectoral girdle of Subterbranchialia is located immediately behind the neurocranium, forcing the branchial chamber under the neurocranium. Gills are covered by an opercular membrane. Anterior vertebrae are either free or fused (synarcuale). The palatoquadrate is free or fused with the neurocranium (autostylic).

Recent Subterbranchialia are represented by the chimaeras, living at depths of 100–1,500 m. Many Paleozoic forms are related to the chimaeras; often only remains of dentition of some forms are known.

FIGURE 78. Elasmobranchii. **a** *Ctenacanthus* (Lower Carboniferous, Ctenacanthoidea), ≈11 cm. **b** Teeth, *Cladodus*-type, width of the large tooth, ≈1.8 cm. **c** *Hybodus* (Middle Triassic–Upper Cretaceous, Hybodontoidea), restoration, up to 2 m or longer. **d** *Strophodus* (Jurassic, Hybodonotidea), tooth, ≈0.5x. **e** *Xenacanthus* (Upper Devonian–Middle Permian, Xenacanthida), restoration, ≈50 cm. **f** *Isurus* (Oligocene, Galeoidea), tooth, ≈0.5x. **g** *Carcharodon* (Pliocene, Galeoidea), tooth, ≈0.5x. **h** *Myliobatis* (Recent, Batoidea), upper teeth, ≈0.2x. **i** *Ptychodus* (Cretaceous, Hybodontoidea), tooth, 0.5x. **k** *Helicoprion* (Carboniferous–Permian, Edestoidea), symphysal teeth, much reduced. These teeth were not discarded. Subterbranchialia. **l** *Sibyrhynchus* (Upper Carboniferous, Iniopterygia), restoration, ≈30 cm. **m** *Ischyodus* (Upper Jurassic, Chimaeroidea), restoration, ≈1.2 m. **n** Thin section of a "bradyodont" tooth of a chimaeroid; rd, tubular dentine, td, trabecular dentine. **o** *Psammodus* (Lower Carboniferous, Petalodontida), dental plates, reduced (after *Heimberg, Jaekel, Moy-Thomas, Zangerl*).

Two types of dentition can be distinguished:

1. Selachoid type: sharklike teeth, periodically replaced
2. Plate type: teeth fused, with no or only slow replacement

Mainly Carboniferous forms such as the bizarre **Iniopterygia** and the **Chondrenchelyida** are now joined to the well-known **Holocephali** sensu stricto. One of the few Paleozoic representatives of the chimaerids is *Helodus* (Upper Carboniferous). Chimaeras were widely distributed during the Jurassic.

3. Class Acanthodii (figure 79)

Acanthodii are documented from as early as the Upper Silurian. They were widely distributed during the Devonian and became extinct during the Lower Permian. They can be characterized as follows:

1. Mostly fusiform fishes generally very small, rarely of larger size (10–20 cm, rarely to 2.5 m). Body covered by small, interlocking scales. Pectoral, pelvic, and one or two dorsal fins, one anal fin, and a heterocercal caudal fin. All fins (except caudal) have an anterior spine. Between pectoral and pelvic fins were one to six pairs of intermediate spines.
2. Jaw suspension hyostylic or amphistylic. Between mandibular and hyoid arches is a spiracle, generally followed by five gill slits, with separate opercula. Dentition absent or composed of mono- or polycuspid teeth and/or dental plates.
3. Notochord persisting, neural and hemal arches ossified.

The origin of Acanthodii is unknown. The opinion that they were related to Chondrichthyes has recently been abandoned in favor of an interpretation that makes them a sister group of the Osteichthyes. All Silurian finds are from marine sediments. By the beginning of the Carboniferous they lived in fresh water as well. The primitive **Climatiida** have three to six pairs of intermediate spines and teeth with many-cusped individual teeth and tooth spirals. Predatory **Ischnacanthida** (Devonian) had strong, toothed jaws. They were generally small, only *Xylacanthus* reached a length of 2.5 m. **Acanthodida** were slim, even eel-shaped, with only one intermediate pair of spines or with none at all. Best known is *Acanthodes bronni* of the European Lower Permian: toothless, with reduced scale cover.

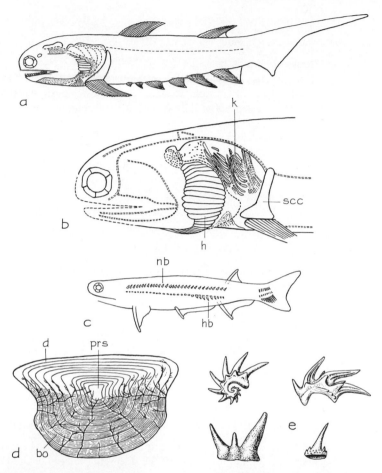

FIGURE 79. Acanthodii. **a** *Climatius* (Upper Silurian–Lower Devonian, Climatiidae), restoration, ≈14 cm. **b** *Euthacanthus* Lower Devonian, Climatiidae), restoration of the head; h, gill operculum of the hyoid arch, k, gill coverings, scc, scapulocoracoid. **c** *Acanthodes* (Lower Carboniferous–Lower Permian, Acanthoididae), restoration, ≈50 cm; hb, hemal arch, nb, neural arch. **d** *Acanthodes,* structure of a scale; bo, bone, d, dentin layers, prs, primordial scale. **e** Teeth of Ischnacanthidae (Silurian–Carboniferous) (after *Denison, Miles, Watson*).

4. Class Osteichthyes (bony fishes)

Important distinguishing characteristics of Osteichthyes are:

1. Body covered by dermal scales.
2. Skeleton usually well ossified. Caudal fin of early forms heterocercal. Dermal bones of the skull show various basic patterns. Dermal components of the pectoral girdle are connected with the skull.
3. Heart with atrium and ventricle, pumps venous blood. Four pairs of aortic arches.
4. Breathing with gills and lungs or gills only.
5. Ten pairs of cranial nerves. Lateral line system.
6. Body temperature variable (poikilothermy).
7. Oviparous, rarely ovoviviparous. Development includes metamorphosis. Brood care may take place.

Osteichthyes encompass two subclasses:

1. Subclass Actinopterygii, ray-finned fishes. Includes most living species.
2. Subclass Sarcopterygii, fleshy-finned fishes. Among them are found ancestors of the Tetrapoda.

The oldest remains of Osteichthyes, scales of Actinopterygii, are known from Upper Silurian. During the Devonian, Actinopterygii and Sarcopterygii were already distinctly separate. Their common ancestors are unknown.

1. Subclass Actinopterygii (ray-finned fishes) (figures 80–82)

Actinopterygii are distinguished from Sarcopterygii by the lack of choanae (inner nasal passages), by the different patterns of the skull

FIGURE 80. Body organization of Actinopterygii. **a** *Cheirolepis* (Devonian, Palaeonisciformes), restoration, ≈70 cm; pc, pectoral fin, pv, ventral (pelvic) fin. **b** Skeleton of a perch (*Lates*, Recent), ≈1 m; m, maxilla, pm, premaxilla, pc, pectoral girdle, pv, pelvic girdle (shifted craniad). **c–e** Shapes of caudal fin: **c** Heterocercal, **d** Homocercal (derived heterocercal, *Amia, Lepisosteus*), **e** Homocercal (Teleostei); h, hypuralia. **f** Section of the dorsal fin of a teleostean; a, actinotrichia (horny fin rays), e, lepidotrichia (bony fin rays). **g, h** Ganoid scale: g, from inside, h, from outside. **i, k** Sections of ganoid scales: **i** *Palaeoniscus* type, **k** *Lepisosteus* type; d, dentine, g, ganoin, k, bone. **l** Elasmoid scale (Teleostean). **n** Cycloid scale (Teleostean) (after *Goodrich, Gross, Lehman, Norman, Starck*).

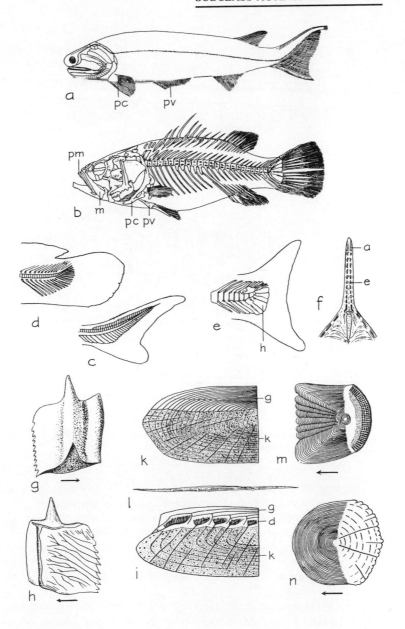

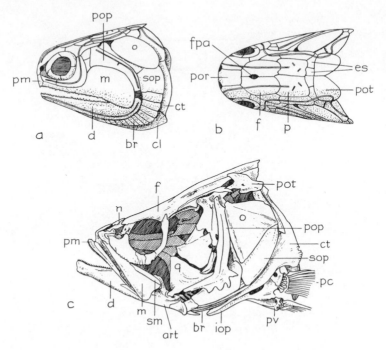

FIGURE 81. Actinopterygii: Skull.
a, b *Moythomasia* (Upper Devonian, Palaeonisciformes). **c** *Lates* (Recent, Teleostei); art, articular, br, branchiostegalia, cl, clavicle, ct, cleithrum, d, dentary, es, extrascapular, f, frontal (parietal), fpa, foramen parietale, iop, interopercular, m, maxilla, n, nasal, o, opercular, p, parietal (postparietal), pc, pectoral fin, pm, premaxilla, pop, preopercular, por, postrostral, pot, posttemporal, pv, pelvic fin, sm, supramaxilla, sop, subopercular (after *Gregory, Jessen*).

components, by possession of ganoid scales, and by the special nature of the endoskeleton of fins. At the beginning of the Mesozoic, Actinopterygii were exclusively marine; the sea remained the main place of their development, though, since the beginning of the Tertiary, numerous forms penetrated into fresh water. Many characteristics of the early representatives became modified during the evolution of the various lines. Table 16 summarizes several developmental trends.

The most noticeable modifications that took place during the Actinopterygii evolution concern the mechanisms of feeding and locomotion. Concurrently, the exoskeleton became lighter and the endoskeleton more heavily ossified.

Even today it is common to lump ancient Actinopterygii with the Chondrostei. However, modern studies of progressive Mesozoic forms and of the Teleostei led to a new grouping of the higher Actinopterygii:

Subclass Actinopterygii

1. Infraclass Chondrostei
2. Infraclass Neopterygii
        Division Ginglymodi (*Lepisosteus*)
        Division Halecostomi
                Subdivision Halecomorphi (*Amia*)
                Subdivision Teleostei

1. Infraclass Chondrostei (figure 82, a–c)

Characteristic representatives are **Palaeonisciformes:** *Cheirolepis* (Devonian), *Palaeoniscus* (Permian). The high-backed **Platysomidei** (Carboniferous–Cretaceous) represent a sideline. Recent Chondrostei are represented by sturgeons (**Acipenseriformes**). The **Polypteriformes,** who have fleshy paired fins and ventral lungs (*Polypterus,* bichir, Africa) are viewed as considerably modified descendants of the Palaeonisciformes.

2. Infraclass Neopterygii (figure 82, d–i)

Division Ginglymodi (figure 82, d–e)

The oldest representative is *Acentrophorus* (Upper Permian). Typical are *Lepidotes* (Triassic–Cretaceous), *Dapedium* (Jurassic), and the recent *Lepisosteus* (gar pike, North America). **Pycnodontiformes** (Triassic–Eocene) are high-backed, with specialized dentition.

Division Halecostomi

Maxilla is freely movable. A conservative (Halecomorphi) and a progressive line (Teleostei) are distinguished.

Subdivision Halecomorphi (figure 82, f)

A well-known representative is *Amia* (Cretaceous–Recent).

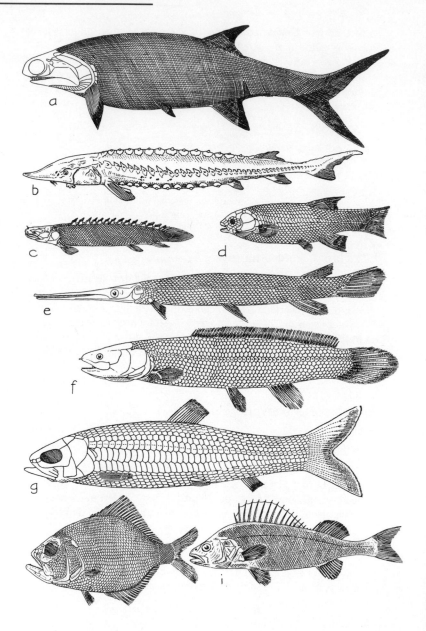

TABLE 16.   Evolutionary Trends of Actinopterygii

Characteristic	Primitive	Modern
Scales: Shape	Rhomboid	Cycloid, ctenoid
Scales: Structure	Ganoin, dentine, isopedine	Isopedine
Endoskeleton	Partially cartilaginous	Ossified
Spiraculum	+	−
Maxilla	Fixed	Free
Clavicle	+	−
Caudal fin	Heterocercal	Homocercal
Ventral fins	Far back	Often shifted forward
Breathing	Gills and lungs	Gills
Swim bladder	−	+

Subdivision Teleostei (figure 82, g–i)

Teleostei show characteristics of modern representatives, as shown in table 16. Their taxonomic position is still under lively discussion. **Pholidophoriformes** (Triassic–Cretaceous) are considered as the central, original group. Two superorders are discussed here:

Superorder Ostariophysi

They possess Weberian ossicles, a row of small bones along each side of the spine, which carry sound waves and changes in pressure from a special chamber of the swimming bladder to the inner ear. This fish group encompasses the majority of recent freshwater fish

FIGURE 82. Actinopterygii. **a–c** Chondrostei: **a** *Pygopterus* (Lower Permian, Palaeonisciformes), ≈40 cm, **b** *Acipenser,* sturgeon (Tertiary–Recent), up to 3 m or longer. **c** *Polypterus,* bichir (Recent, Polypteriformes), ≈50 cm. **d–i** Neopterygii. Ginglymodi: **d** *Acentrophorus* (Upper Permian, Semionotiformes) ≈7 cm, **e** *Lepisosteus,* gar pike (Tertiary–Recent, Lepisosteiformes), up to 3 m. Halecomorphi: **f** *Amia,* bowfin (Upper Cretaceous–Recent, Amiiformes), up to 60 cm. Teleostei: **g** *Pholidophorus* (Jurassic, Pholidophoriformes) ≈15 cm. **h** *Berycopsis* (Cretaceous, Beryciformes), a primitive acanthopterygian, ≈30 cm, **i** *Perca,* perch (Eocene–Recent, Perciformes), ≈20 cm (after *Bertin, Gill, Lehman, Moy-Thomas, Rayner, Smith-Woodward*).

fauna. Examples are **Cyprinidae** (carps and relatives) with pharyngeal teeth, and **Siluridae** (catfishes).

Superorder Acanthopterygii (spiny-rayed fishes)

The Acanthopterygii encompass the majority of modern marine fishes. Dorsal and anal fins with spines. The largest order of all vertebrates, **Perciformes** (perches), with more than 6,000 species, belongs to this superorder. Their radiation occurred during the Tertiary.

2. Subclass Sarcopterygii (fleshy-finned fishes)

Osteichthyes whose paired fins have a central endoskeleton with a fleshy base covered by scales. Dermal bones, scales, and lepidotrichia follow the cosmoid pattern (figure 84, e). Breathing is accomplished with lungs and gills. Two well-delineated groups, namely, Crossopterygii and Dipnoi (lungfishes), whose common ancestors lived during the Silurian or earlier, can be distinguished, even among the oldest known Sarcopterygii of the Lower Devonian. During Middle Devonian specialized Crossopterygii appeared: Actinistia and Onychodontida. The ancestors of tetrapods are found among the Crossopterygii.

1. Infraclass Dipnoi (lungfishes) (figure 83)

Long-lived group of Sarcopterygii; during the course of evolution, highly specialized. Their developmental trends are summarized in table 17.

The skull is a mosaic of numerous dermal bones; homology with those of Crossopterygii is impossible to discern. Their rapid specialization occurred in early geological periods. Dipnoi flourished during the Upper Devonian and the Carboniferous. Many cannot as yet be classified in a proper phylogenetic system. *Dipterus* (Devonian) is a well-known representative; in appearance it resembles a crossopterygian. A cosmopolitan distribution had *Ceratodus* (Triassic–Cretaceous), a possible predecessor of the Australian *Neoceratodus*. The recent Lepidosirenidae represent an evolutionary sideline.

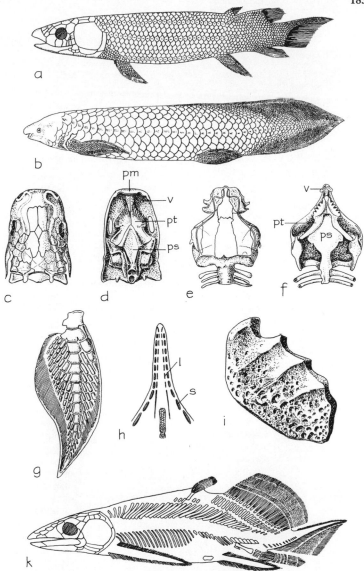

FIGURE 83. Dipnoi. **a** *Dipterus* (Devonian), restoration, ≈40 cm. **b** *Neoceratodus* (Recent), up to 1.8 m. **c, d** *Dipterus,* skull, **c** Dorsal view, **d** Ventral view; **e, f** *Neoceratodus,* skull, **e** Dorsal view, **f** Ventral view; pm, premaxilla, ps, parasphenoid, pt, pterygoid, v, vomer (pt and v carry teeth). **g** Pectoral fin of *Neoceratodus*. **h** *Dipterus*, sectional view of dorsal fin; l Lepidotrichia (bony fin rays), s, scale. **i** *Ceratodus* (Triassic), tooth. **k** *Fleurantia* (Upper Devonian), restoration, ≈24 cm (after *Forster-Cooper, Goodrich, Graham-Smith* and *Westoll, Jarvik, Romer, Starck*).

TABLE 17.   Evolutionary Trends of Dipnoi

Characteristic	Devonian Forms	Modern Forms
Number of dermal bones of the skull	Very large	Small
Foramen parietale	+	−
Endocranium	Ossified	Cartilaginous
Scales	Cosmoid, thick	Without cosmine, thin
Caudal fin	Heterocercal	Gephyrocercal
Median fins	2 dorsal, 1 anal	Continuous with the caudal fin
Fin support	Lepidotrichia	Ceratotrichia
Dentition	Numerous denticles	Teeth fused into a few dental plates

List of common abbreviations used for the description of skull elements

a	angular	po	postorbital
art	articular	pop	preopercular
br	branchiostegalia	por	postrostral
d	dentary	pos	postsplenial
ec	ectopterygoid	pot	posttemporal
es	extrascapular	pp	postparietal (parietal)
esl	lateral	prf	prefrontal
esm	median extrascapular	ps	parasphenoid
f	frontal (nasal)	pt	pterygoid
hy	hyomandibular	q	quadrate
iop	interopercular	qj	quadratojugal
it	intertemporal	s	splenial
j	jugal	sa	surangular
l	lacrimal	sm	supramaxilla
m	maxilla	so	suborbital
n	nasal	soc	supraoccipital
o	opercular	sop	subopercular
p	parietal (frontal)	sp	spiracular
pf	postfrontal	sq	squamosal
pl	palatine	st	supratemporal
pm	premaxilla	t	tabular
pn	postnarial	v	vomer

FIGURE 84. Crossopterygii. **a–c** Skull, **a** Dorsal view, **b** Ventral view, **c** Lateral view. **d** Neurocranium: 1, ethmosphenoid segment, 2, otico-occipital segment, movably jointed. **e** Structure of the exoskeleton (cosmoid type); cos, cosmine (d, dentine, e, enameloid), k, cancellous bone, is, isopedine (bonelike material) (after *Bystrow, Romer*).

Infraclass Crossopterygii (figures 84, 85)

Crossopterygii have a neurokinetic skull. It consists of an anterior and a posterior segment, connected by a joint that allows lifting of the anterior part. Crossopterygii also have a functional basipterygoid joint.

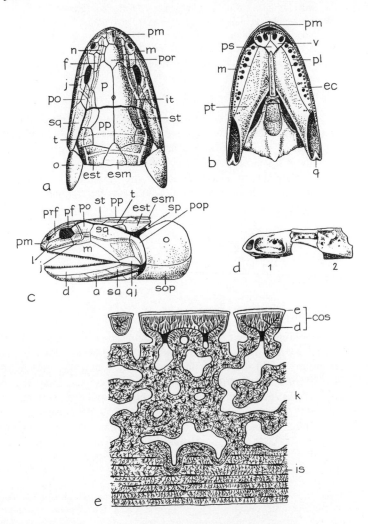

Five orders belong to Crossopterygii; their interrelationships are currently under discussion. Actinistia are often afforded a special status.

Infraclass Crossopterygii

1. Order Actinistia (Coelacanthiformes), Devonian–Recent
2. Order Holoptychiida (Porolepiformes), Devonian
3. Order Osteolepididia (Osteolepiformes), Devonian–Permian
4. Order Rhizodontida, Carboniferous
5. Order Onychodontida (Struniiformes), Devonian

1. Order Actinistia (figure 85, a, b)

Actinistia lack choanae and foramen parietale. Characteristic is the diphycercal caudal fin with an additional central lobe. In contrast to the generally small fossil representatives, the living *Latimeria* reaches a length of 1.8 m. This predatory fish lives in the sea, 150–800 m deep, in the Comoro Islands region of the Indian Ocean. It breathes with gills; the rudimentary lung is filled with fatty connective tissue. It is ovoviviparous.

2. Order Holoptychiida (figure 85, c)

There are only a few, mostly incompletely known genera from the Devonian.

3. Order Osteolepididia (figure 85, d, e)

Osteolepididia possess true choanae; many have a foramen parietale as well. Dermal pectoral girdle with an interclavicle, clavicle, and cleithrum. Paired fins are usually roundish, with a uniserial archipterygium. The homology of dermal skull components has long been disputed. Currently it is believed that the components, formerly designated "frontalia," are in fact parietalia. The oldest Osteolepididia were marine. *Eusthenopteron* (Upper Devonian), a freshwater form, is by far the best known of all fossil fishes. It is among the Osteolepididia that the ancestors of tetrapods are sought.

FIGURE 85. Crossopterygii. **a** *Latimeria* (Recent, Actinistia), up to 1.8 m. **b** *Laugia* (Triassic, Actinistia), restoration ≈50 cm. **c** *Holoptychius* (Devonian, Holoptychiida), restoration, ≈50 cm. **d, e** *Eusthenopteron* (Upper Devonian, Osteolepididia), restoration, ≈60 cm. **f** *Strunius* (Devonian, Onychodontida), restoration, ≈6 cm (after *Jarvik, Jessen, Lehman*).

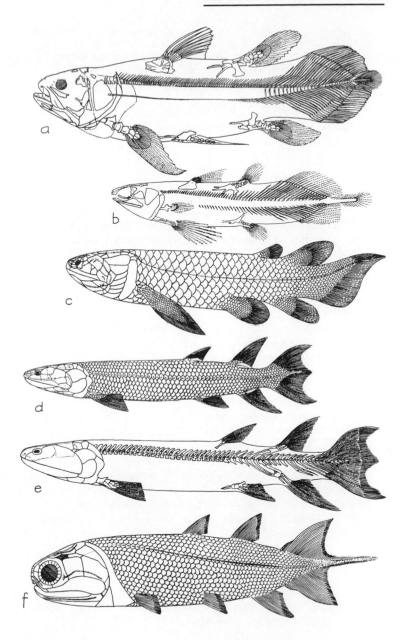

4. Order Rhizodontida

Small order of Devonian freshwater fishes; extinct during the Carboniferous. Not in the mainstream of tetrapod evolution.

5. Order Onychodontida (figure 85, f)

Only two genera are known to date, *Strunius* and *Onychodus,* Devonian.

Conquest of the Land—The Tetrapods

The change from an aquatic to a terrestrial way of life is the most important step of the vertebrate evolution. Although the fishes have an enormous habitat at their disposal—the sea alone covers about 70% of the earth's surface—the water never did stimulate the vertebrate evolution to the extent that the land did. Changes that an animal must undergo to accomplish the transition from aquatic to terrestrial life concern the entire organism. Every amphibian generation repeats to some extent, in the course of a few weeks, the same metamorphosis that took place during thousands of centuries of phylogeny. The most important modifications concern the organs of breathing and locomotion. Part of the branchial arch was modified to allow the conversion from gill to lung breathing. Even the young of higher tetrapods, who do not depend on gills, develop blood vessel sections that show how the blood supply of head and neck was based on the branchial arches. Branchial arteries are laid out in humans as well—an important testimony to the unified structural plan of all vertebrates. Changes to the apparatus of locomotion are equally profound. Greatest demands are made on the spinal column and on the paired extremities. A firm attachment of the appendages to the spinal column is required; it is achieved through the development of shoulder and pelvic girdles. While the spinal column of fishes has only two sections (thoracic and caudal), the tetrapod spine has four: cervical, thoracic, sacral, and caudal (figure 86, a–f).

AXIAL SKELETON (figure 86, g–o)

The type of vertebrae is very important for the classification of lower tetrapods. Two basic types can be distinguished in the oldest tetrapods: apsispondylic and lepospondylic.

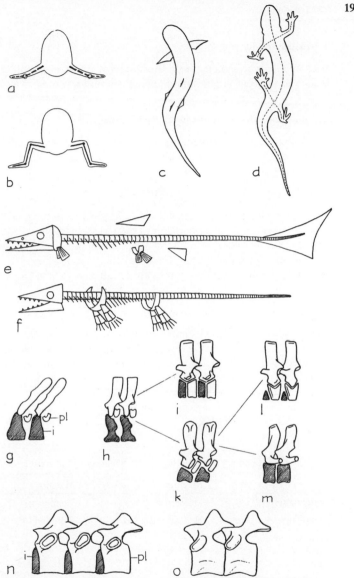

FIGURE 86. Conquest of the land. **a, b** Cross section of the thoracic region of a fish (**a**) and a primitive tetrapod (**b**). **c** Swimming motion of a fish. **d** Crawling and slithering of a primitive tetrapod. **e** Diagram of a fish skeleton. **f** Diagram of a tetrapod skeleton. **g–m** Apsispondylic vertebrae: **g** *Eusthenopteron* (Crossopterygii), **h** *Ichthyostega* (the oldest known tetrapod), **i** Embolomerous vertebral type, **k** Rhachitomous type, **l** Primitive reptilian type, **m** Stereospondylous type; i, intercentrum, pl, pleurocentrum. **n, o** Lepospondylous vertebrae (after *Carroll, Colbert, Kuhn-Schnyder, Peters* and *Gutmann, Starck*).

1. Apsispondylic vertebra

It can be derived directly from the vertebrae of the crossopterygians; occurs in the fossil Labyrinthoidea and (modified) in the Amniota.

2. Lepospondylic vertebra (socket vertebra)

Socketlike amphicoelic vertebra such as is found both in small Paleozoic and the modern amphibians.

SKELETON OF THE GIRDLES (figure 87, a–d)

The endoskeletal girdles (scapulocoracoid, pelvis) of tetrapods are much strengthened in comparison to the Crossopterygii. The dermal bones of the shoulder girdle become more important. Both halves of the shoulder girdle are firmly linked by the interclavicle. The enlarged, massive pelvis becomes firmly attached to the spine.

SKELETON OF THE APPENDAGES (figure 87, e–g)

Paired fins of the fishes (ichthyopterygia) are homologous with the limbs of tetrapods (chiropterygia). With the transition from aquatic to terrestrial life, the skeleton of fins was transformed into a system of levers. Terms used for the equivalent parts of the appendages, both anterior and posterior, are summarized in table 18.

The basipodium consists of numerous bones that originally had equivalent patterns in both the carpus and tarsus (figure 87, f, g).

TABLE 18.   Terminology Relating to the Skeleton of the Extremities

	Forelimb	Hindlimb
Stylopodium	Humerus	Femur
Zeugopodium	Radius and Ulna	Tibia and Fibula
Autopodium (hand = manus, foot = pes) primarily 5-rayed		
Basipodium	Carpus	Tarsus
Metapodium	Metacarpus	Metatarsus
Acropodium	Phalanges	Phalanges

TABLE 19.   Terminology Relating to the Components of Carpus and Tarsus

Carpus of Primitive Tetrapods	Mammalia	Tarsus of Primitive Tetrapods	Mammalia
Radiale	Fused with radius	Tibiale	Fused with tibia
Intermedium	Lunate	Intermedium	Astragalus = Talus
Ulnare	Triquetrum	Fibulare	Calcaneus
Centralia 1–4	Navicular = Scaphoid; Central	Centralia 1–4	Navicular = Scaphoid (partially retained in Talus)
—	Pisiform	—	Pisiform fused with Calcaneus
Carpal 1	Trapezium	Tarsal 1	Cuneiform 1
Carpal 2	Trapezoid	Tarsal 2	Cuneiform 2
Carpal 3	Capitatum	Tarsal 3	Cuneiform 3
Carpal 4	Hamatum	Tarsal 4	Cuboid
Carpal 5	Rudimentary	Tarsal 5	Rudimentary

An overview of the terminology of tarsal and carpal bones is given in table 19.

The number of the phalanges is variable. Primitive reptiles had phalangeal formulae 2 3 4 5 3 (or 4), counting from ray I. Amphibians usually have fewer phalanges.

Front and rear extremities of primitive tetrapods are spread out, because the humerus and femur are horizontal. Radius and ulna, or tibia and fibula, are bent at a right angle in the elbow and knee joints; the hand and foot point forward. Humerus and femur later remain closer in, positioning the hand and foot more under the body.

SKULL COMPONENTS (figure 88)

The tetrapod skeleton encompasses both exoskeletal (dermal bones) and endoskeletal components (cartilage and replacement—endochondral—bone). Functionally, three main units can be distinguished on a primitive skull:

1. Dermal Roof of the Skull (of exoskeletal origin only)

The roof has openings for the nose (nares), eyes (orbitae), and the median eye (foramen parietale or parietal foramen). Ear notches (for the eardrum openings) are found on both sides of the posterior edge

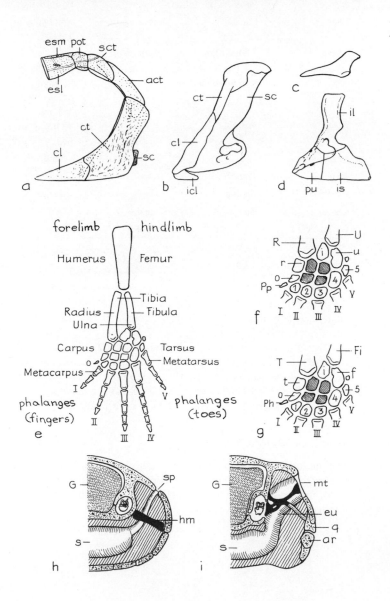

in place of the spiracle of fishes. Dermal bone components of the skull roof are as follows (with their usual abbreviations):

a) Tooth bearing components of the edge: premaxilla = pm, maxilla = m.
b) Paired components along the median line: nasal = n, frontal = f, parietal = p, postparietal = pp. In addition to the nasal, a septomaxilla (= sm) may also occur.
c) Paired components in the orbital region: prefrontal = prf, postfrontal = pf, postorbital = po, jugal = j, lacrimal = l.
d) Paired temporal row: intertemporal = it, supratemporal = st, tabular = t.
e) Dermal bones of the quadrate: quadratojugal = qj, squamosal = sq.

2. Palatal Complex (of endochondral and dermal bone)

Endochondral bones of the palatoquadrate: epipterygoid = e, quadrate = q
Dermal bones: pterygoid = pt, vomer = v, palatine = pl, ectopterygoid = ec.

The palate is divided by an interpterygoid cleft of varying size. Another dermal bone is visible in the cleft: parasphenoid = ps. It may join with the pterygoids near the basipterygoid joint.

3. Neurocranium (endochondral bone)

Occiput: exoccipital (paired) = ec, supraoccipital = soc, basioccipital = bo.
Otic capsule: prootic = pr, opisthotic = op.
Basis of the neurocranium: basisphenoid = bs, with pituitary fossa.

FIGURE 87. Primitive tetrapods. **a, b** Pectoral (shoulder) girdle, **a** *Eusthenopteron* (Crossopterygii), **b** *Eryops* (primitive tetrapod); act, anocleithrum, cl, clavicle, ct, cleithrum, esl, lateral extrascapular, esm, medial extrascapular, icl, interclavicle, pot, posttemporal, sc, scapula, sct, supracleithrum. **c, d** Pelvis of a fish (**c**) and a primitive tetrapode (**d**); il, ilium, is, ischium, pu, pubis. **e** Schematic of a primitive tetrapode extremity. **f** Schematic of carpus of a primitive tetrapode; R, radius, U, ulna, r, radial, i, intermedium, u, ulnar, 1–5, distal carpalia, Pp, praepollex (ray 0), I–V, metacarpalia I–V. The four centralia are cross-hatched. **g** Diagram of tarsus of a primitive tetrapode; T, tibia, Fi, fibula, t, tibial, 1–5, distal tarsalia 1–5, Ph, prehallux (ray 0), I–V, metatarsalia I–V. The four centralia are cross-hatched. **h, i** Development of the middle ear and the auditory ossicles of a fish (**h**) and a primitive tetrapode (**i**) (Amphibium); ar, articular, eu, eustachian tube, G, brain, hm, hyomandibular, mt, eardrum, q, quadrate, s, pharynx, sp, spiracle (after *Jarvick, Romer*).

The lateral walls of the cranium are not ossified in the early tetrapods.

Components of the Lower Jaw (endo- and exoskeletal)
(figure 88, e, f)

Endoskeleton: Meckel's cartilage, endochondral bone: articular = ar. Exoskeleton (dermal bone): dental = d, splenial = sp, angular = a, surangular = sa, coronoid = cor, prearticular = pra.

During the phylogeny, the tetrapod skull had undergone numerous changes, primarily concerning altered proportions, reduction of the components of the dermal roof, occurrence of additional openings, loss of foramen parietale, and modification of the palate.

5. Class Amphibia (amphibians)

Recent representatives of the class are frogs and toads (Anura), salamanders and newts (Urodela), and caecilians (Apoda). Zoologists group them as Lissamphibia. In addition, there is a great variety of fossil forms that so far cannot be genetically linked with the Lissamphibia. The following are amphibian characteristics (figure 89):

1. Skin with horny scales or naked and moist with many glands.
2. Skeleton considerably ossified. Skull autostylic, originally kinetic. Dermatocranium closed or reduced. One or two occipital condyls. Quadrato-articular joint. One auditory ossicle (columella = stapes).
3. Vertebral body has various shapes. Generally only one sacral vertebra. Shoulder girdle not connected to the cranium. Two pairs of originally pentadactylous limbs.
4. Heart with one ventricle and two atria; four pairs of aortic arches.
5. Lung breathing, also through the skin, mouth cavity, or gills, separately or in combination.
6. Brain with ten or twelve pairs of cranial nerves.
7. Variable body temperature (poikilothermy).
8. Fertilization external or internal, oviparity or, rarely, ovoviviparity. Development includes metamorphosis, usually in fresh water. No embryonic sheaths (Anamniota). Larvae with gills. Brood care may occur.

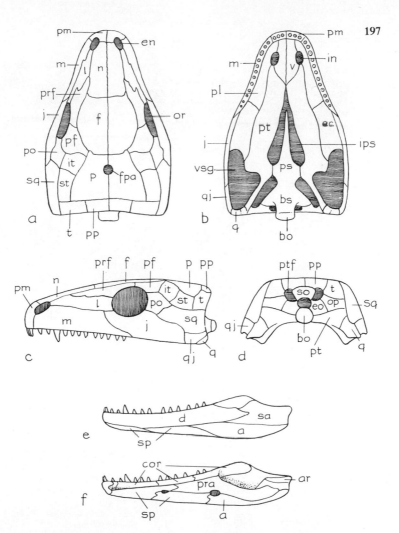

FIGURE 88. Schematic of a primitive tetra-pod skull. **a** Dorsal view; **b** Ventral view; **c** Lateral view; **d** From behind. Lower jaw; **e** From the outside; **f** From the inside. a, angular, ar, articular, bo, basioccipital, bs, basisphenoid, cor, coronoid, d, dentary, do, dermal supra-occipital, ec, ectopterygoid, eo, exoc-cipital, f, frontal, it, intertemporal, j, jugal, l, lacrimal, m, maxilla, n, nasal, op, opisthoticum, or, orbita, p, pari-etal, pf, postfrontal, pl, palatine, pm, premaxilla, po, postorbital, pp, postparietal, pr, prooticum, pra, prearticular, prf, prefrontal, ps, para-sphenoid, pt, pterygoid, ptf, post-temporal fenestra, q, quadrate, qj, quadratojugal, sa, surangular, so, supraoccipital, sp, splenial, sq, squa-mosal, st, supratemporal, t, tabular, v, vomer. Openings: en, outer nares open-ing, fpa, foramen parietale, in, inner nares opening, ips, interpterygoid cleft, vsg, subtemporal fenestra.

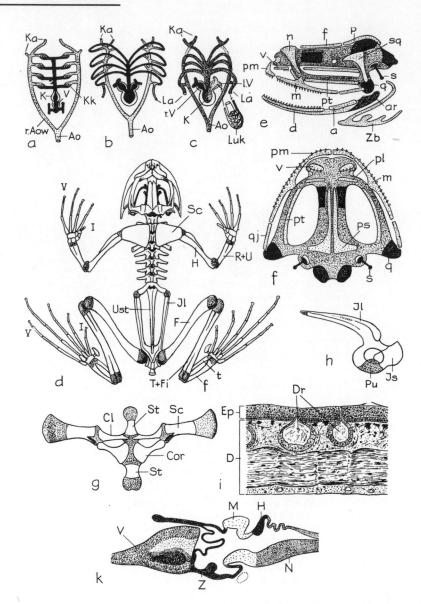

Distribution: Devonian–Recent.

All the important sensory organs of the amphibians were already developed in fishes. Amphibians are the only tetrapod group that retains the lateral line. Their brain hemispheres are still primarily olfactory centers. The Jacobson's organ is first documented in amphibians; it is an invagination inside the nasal passage, serving for chemical testing of food.

Amphibians first appeared in late Devonian; the Carboniferous and the Permian were their main periods of prosperity, followed by a decline at the end of the Triassic. A new radiation occurred during the Jurassic, already involving representatives of the modern amphibians (figure 90). In many respects, both the form and biology of the early amphibians were extraordinarily different from what they are in the modern ones. The shape of the vertebral body is important for classification of the early amphibians.

Class Amphibia

1. Subclass Labyrinthodontia (mostly large), Upper Devonian–Triassic
2. Subclass Lepospondyli (mostly small), Carboniferous–Permian
3. Subclass Lissamphibia (modern representatives and their ancestors), Triassic–Recent

Lepospondyli and Lissamphibia are probably not natural units.

1. Subclass Labyrinthodontia

Representatives of the first amphibian radiation, encompassing practically all of the large Paleozoic and Triassic forms. Apsispondylic vertebrae, body mostly covered with scales. The pattern of dermato-

FIGURE 89. Body plan of Amphibia.
a–c Heart and aortic arches in fishes (a), amphibian larvae (b), and in Urodela (c). Ao, aorta, Aow, aortal stem, K, ventricle, Ka, carotid, Kk, gill capillaries, La, lung artery, Luk, lung capillaries, V, atrium. d Skeleton of a frog; F, femur, H, humerus, Il, ilium, R + U, radius/ulna, Sc, scapula, T + Fi, tibia/fibula, t, tibiale, f, fibulare, Ust, urostyle (fused caudal vertebrae). e, f Skull of a frog; Zb, hyoid. g Pectoral girdle of a frog; Cl, clavicle, Cor, coracoid, Sc, scapula, St, sternum. h Pelvis of a frog; Il, ilium, Is, ischium, Pu, pubis (cartilaginous). i Amphibian skin in cross section; D, dermis, Dr, glands, Ep, epidermis. k Frog brain in longitudinal section; H, metencephalon, N, myelencephalon, V, prosencephalon, Z, diencephalon. For other abbreviations, see figure 88. (after *Gaupp, Hildebrand, Kühn, Parker* and *Haswell*).

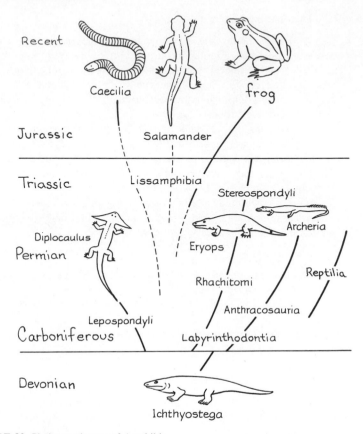

FIGURE 90. Phylogenetic tree of Amphibia.

cranium, labyrinthodont teeth, vertebrae, and girdles of some early representatives were similar to those of certain crossopterygians. Besides the oldest known tetrapods (Ichthyostegalia), two other orders can be distinguished: Temnospondyli and Anthracosauria.

1. Order Ichthyostegalia (figure 91)

The most primitive of all known amphibians, from the Upper Devonian. They show a mixture of fishlike and amphibian characteristics (table 20).

Finds from Greenland and Australia testify to a wide distribution of the Ichthyostegalia. They cannot, however, be considered direct ancestors of the other Labyrinthoidea.

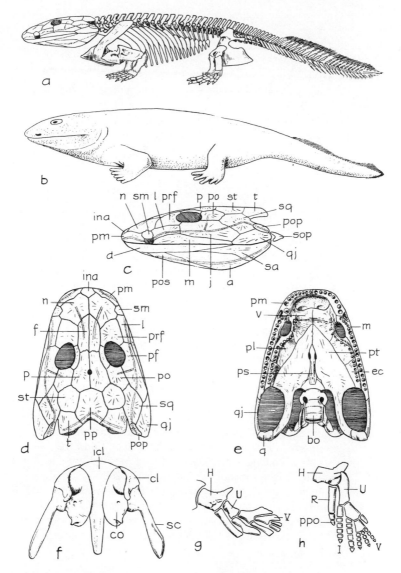

FIGURE 91. Labyrinthodontia: Ichthyostegalia, *Ichthyostega* (Upper Devonian, Greenland). **a** Restoration of skeleton, **b** Appearance of a live animal, ≈1 m long, **c–e** Skull; ina, internasal, pop, preopercular, pos, postsplenial, sop, subopercular, **f** Shoulder girdle; co, coracoid. **g** *Eusthenopteron* (Upper Devonian, Crossopterygii), pectoral fin. **h** *Ichthyostega,* forelimb. Other abbreviations as in figures **87, 88.** (after *Jarvik*).

TABLE 20.   Characteristics of Ichthyostegalia

Fish Characteristics	Amphibian Characteristics
Median fin with radialia and lepidotrichia	Pentadactylous limbs
	Bipartite ribs
Rudimentary preopercularia and subopercularia	Pectoral girdle no longer connected to the skull
Lateral line system in closed canals	Skull anteriorly short, posteriorly large
Traces of the intracranial joint	
Ethmosphenoid ossified	Undivided dermatocranium
Short parasphenoid	

2. Order Temnospondyli (figures 92, 93)

A large group of Labyrinthodontia (Carboniferous–Permian) with a characteristic pattern of the dermatocranium. The small tabular is separated from the parietal by the supratemporal. An otic notch is common. The vertebral body shows a tendency to strengthening of the intercentrum at the expense of the pleurocentra. The skull, originally fairly high, becomes more and more flattened, the palate develops a large cleft, the basipterygoid joint disappears, and the one-headed occipital condyle is doubled. Those forms with both pleurocentra and intercentra are grouped together as **Rhachitomi:** *Loxomma* (Carboniferous), *Eryops* (Carboniferous–Permian), *Edops* (Permian). Forms with only one intercentrum are called **Stereospondyli.** In addition to terrestrial forms, there were semiaquatic, secondary aquatic, and neotenic forms, short-snouted and long-snouted types. Temnospondyli were the dominant terrestrial vertebrates during the Carboniferous. During the Permian they began to decline, while reptiles became dominant. During late Permian and during the Triassic, the aquatic Stereospondyli were more abundant. Only a single amphibian group, **Trematosauridae** (Triassic), penetrated into the marine habitat. Among the most aberrant forms of the Triassic belonged **Plagiosauridae.** The skull of *Gerrothorax* is extremely flat and broad. These animals kept larval characteristics throughout their life (neoteny).

3. Order Anthracosauria (figure 94)

A small group of Labyrinthodontia that never survived the Permian/ Triassic transition. The skull has tabular bordering on the parietal,

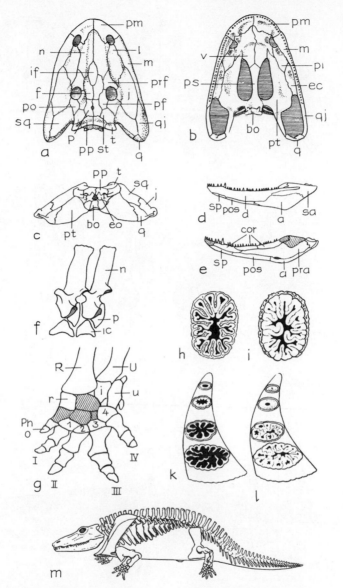

FIGURE 92. Labyrinthodontia: Temnospondyli. **a–g** *Eryops* (Lower Permian, Rhachitomi), **a–e** Skull and lower jaw; pos, postsplenial, **f** Vertebra (rhachitomous type); ic, intercentrum, n, neural arch, p, pleurocentrum, **g** Hand (manus). **h–l** Labyrinthodont tooth structure: **h** *Eusthenopteron* (Crossopterygii), **i** *Benthosuchus* (Lower Triassic, Labyrinthodontia), **k, l** *Benthosuchus*, structure of a young and an old tooth. **m** *Eryops*, restoration of skeleton, ≈1.5 m. Other abbreviations as in figures **87, 88.** (after *Bystrow, Gregory, Romer*).

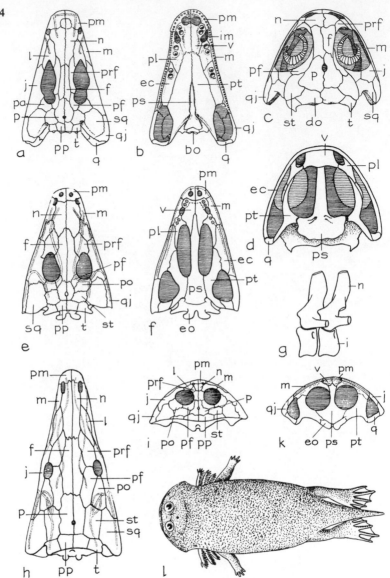

FIGURE 93. Labyrinthodontia: Temnospondyli. **a, b** *Megalocephalus* (Permian, Rhachitomi), skull, dorsal (a) and ventral views (b), ≈35 cm. **c, d** *Branchiosaurus* (Upper Carboniferous, Rhachitomi), skull from above (c) and below (d), ≈1.6 cm. **e–g** *Mastodonsaurus* (Triassic, Stereospondyli), skull, dorsal view (e), below (f), more than 1 m long. **g** Stereospondylous vertebrae; i, intercentrum, n, neural arch. **h** *Trematosaurus* (Lower Triassic, Rhachitomi), skull, dorsal view, ≈20 cm. **i–l** *Gerrothorax* (Upper Triassic, Plagiosauria), skull in dorsal (i) and ventral views (k), l Appearance of a live animal, ≈90 cm. Other abbreviations as in figure **88** (after *Beaumont, Boy, Bystrow, Fraas, Nilsson*).

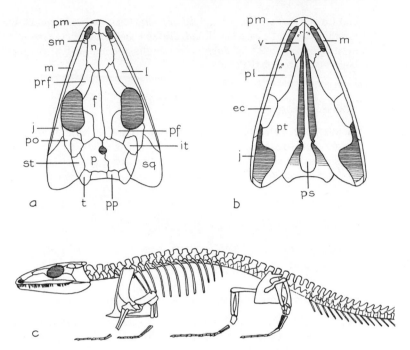

FIGURE 94. Labyrinthodontia: Anthracosauria, *Brukterpeton* (Upper Carboniferous, Gephyrostegida), **a, b** Skull in dorsal (**a**) and ventral (**b**) views, **c** Restoration of the skeleton, ≈17 cm long. Abbreviations as in figure **88** (after *Boy*).

otic notch, and a basipterygoid joint. The vertebral body shows a tendency to develop pleurocentra and intercentra equally (embolomeric type) or to build up pleurocentra at the expense of the intercentrum. The ancestors of Reptilia are sought among the Anthracosauria. Anthracosauria (Lower Carboniferous–Permian) encompass three important evolutionary lines: (1) large to very large **Embolomeri** had an elongated body and were aquatic; (2) the small **Gephyrostegida** of the Carboniferous had nearly reptilian body proportions and are considered to be "models" of the reptile ancestors; and (3) **Seymouriamorpha** (Permian) had the greatest variety of forms. *Seymouria* was for a long time considered to be a reptile; however, skull characteristics and stratigraphic studies argue against such a classification.

Various other groups that lived during the Upper Carboniferous and Lower Permian have amphibian characteristics but approach the reptilian level of evolution. **Diadectidae** (Upper Carboniferous–Permian) are a group of large, herbivorous amphibians, with a reptilianlike postcranial skeleton.

2. Subclass Lepospondyli (figure 95)

Mostly small Paleozoic amphibians with teeth that lack the labyrinthodont structure. Skull without otic notch. The vertebral body consisted of one bony cylinder (socket vertebra), possibly equivalent to the pleurocentra. The following orders do not represent a natural unit.

1. Order Aistopoda (Carboniferous–Permian)

Aquatic Lepospondyli with snake-shaped bodies, with up to 220 vertebrae. Appendages and girdles were secondarily reduced.

2. Order Nectridia (Upper Carboniferous–Upper Permian)

Lepospondyli with characteristically built neural and hemal arches. *Scincosaurus* (Upper Carboniferous), terrestrial. Skull of the aquatic *Diplocaulus* (Upper Permian) is short and broad, with massive tabular horns.

3. Order Microsauria (Upper Carboniferous–Permian)

A group with a wide variety of forms, containing numerous scaly, carnivorous genera. The key characteristic is considered to be the articulation between the skull and the first cervical vertebra. Families **Lysorophidae** and **Adelogyrinidae** are often added to the Microsauria.

FIGURE 95. Lepospondyli. **a–c** Aistopoda. **a** *Dolichosoma* (Upper Carboniferous), ≈70 cm. **b, c** *Phlegethontia* (Upper Carboniferous–Lower Permian), skull, lateral view (**b**), dorsal view (**c**), soc, supraoccipital, ≈1 cm. **d–g** Nectridia; **d–f** *Urocordylus* (Upper Carboniferous), **d** Restoration of skeleton, ≈50 cm, **e** Dorsal, sacral, and two caudal vertebrae, **f** Skull, dorsal view. **g** *Diplocaulus* (Lower Permian), dorsal view of the skull, width ≈30 cm. **h–m** Microsauris: *Pantylus* (Lower Permian), **h, i** Skull in dorsal (**h**) and ventral (**i**) views, **k** Manus, l Pes (foot), **m** Restoration of the skeleton, approx. life-size. For abbreviations see figures **87, 88.** (after *Carroll, Fritsch, McGinnis, Steen, Williston*).

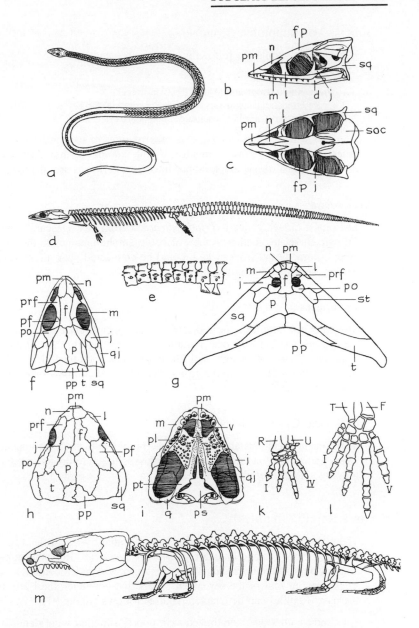

3. Subclass Lissamphibia (figure 96)

Herpetologists unite the following three orders under the Lissamphibia:

1. Order Anura (frogs and toads), Triassic–Recent
2. Order Urodela (salamanders and newts), Jurassic–Recent
3. Order Apoda (Caecilia, caecilians), Paleocene–Recent

The structure and biology of representatives of the three living orders differ widely, just as much from one another as they do from Paleozoic and Triassic amphibians. Intermediate forms are unknown.

1. Order Anura

The oldest known genus is *Triadobatrachus* from Madagascar, Lower Triassic. It relates neither to the Paleozoic amphibians nor to the other Anura. Fossil frogs, which could be connected with the modern genera, appeared during the Jurassic.

2. Order Urodela

Most modern families appeared during the Upper Cretaceous, and many genera in the Miocene. Famous is *Andrias scheuchzeri* of the Oeningen (Baden) Miocene; in 1726 it was described by J. J. SCHEUCHZER as Homo diluvii testis, that is, "skeleton of a man drowned in the deluge."

3. Order Apoda (Caecilia, Gymnophiona)

Most recent caecilians are limited to the humid tropics. Their appendages and girdles had disappeared. As fossils they are documented only by a single vertebra (Paleocene).

6. Class Reptilia (reptiles)

The class Reptilia is in modern times represented by the turtles (Testudinae), New Zealand tuatara (*Sphenodon punctatus*, Rhynchocephalia), lizards and snakes (Squamata), and crocodiles (Crocodilia). These are but four of the seventeen orders that populated the Earth in the past. Reptiles can be characterized as follows (figure 97):

1. Body with a dry, keratinized skin poorly supplied with glands, covered by horny scales, shields, or "hairs."

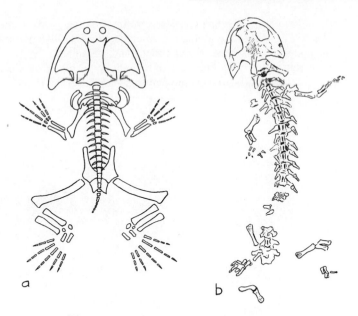

FIGURE 96. Lissamphibia. **a** *Triadobatra-chus* (*Protobatrachus*, Triassic), oldest known anuran. Restoration of skeleton, ≈10 cm long. **b** *Andrias* (Miocene), giant salamander, ≈0.1x (after *Piveteau, Kuhn-Schnyder*).

2. Skeleton ossified. Skull with or without temporal fenestrae, usually only one main condyle. Quadrato-articular joint, one auditory ossicle (columella = stapes).
3. Atlas-axis complex. Differentiation of a neck. Two or more sacral vertebrae. The vertebral body equivalent to the pleurocentrum. Two pairs of originally pentadactylic appendages, phalangeal formula 2 3 4 5 3 (4), distal phalanges clawed.
4. Heart has two atria, one incompletely divided ventricle, and three arterial stems.
5. Lung breathing.
6. Twelve pairs of cranial nerves.
7. Variable body temperature (poikilothermy).
8. Internal fertilization. Eggs generally large, rich in yolk, shelled. Usually oviparous, rarely ovoviviparous. Embryonic membranes: amnion, chorion, allantois (Amniota). Development without metamorphosis. Brood care may occur.

Distribution: Carboniferous–Recent.

Among vertebrates, reptiles were the first real lords of the Earth. Table 21 shows the important characteristics of the living reptiles in comparison with those of amphibians and mammals.

Reptiles exhibit a mixture of ancient and progressive characteristics. Special phylogenetic significance is ascribed to the presence

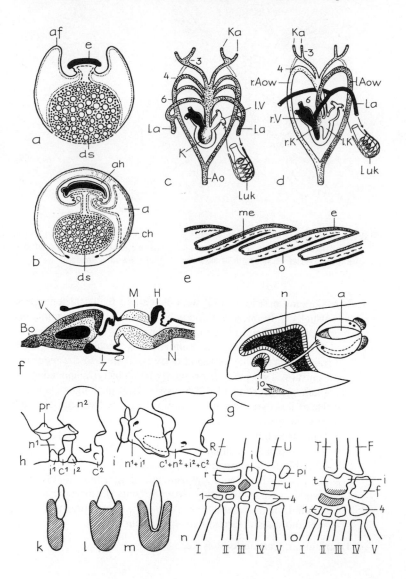

or absence of the temporal fenestrae (figure 98). The following established terms describe their number and position:

no temporal fenestra	anapsid
one, low situated fenestra	synapsid
one, dorsally situated fenestra	euryapsid
two temporal fenestrae	diapsid

However, as the occurrence and position of the temporal fenestrae are the result of mechanical factors, their significance ought not be overrated. The classification of reptiles shown below takes further features into consideration as well.

Class Reptilia

1. Subclass Anapsida (anapsid)
2. Subclass Chelonomorpha (modified anapsid)
3. Subclass Eosuchiamorpha (diapsid)
4. Subclass Lepidosauria (diapsid, modified diapsid)
5. Subclass Archosauria (diapsid)
6. Subclass Pterosauromorpha (diapsid)
7. Subclass Sauropterygomorpha (modified diapsid)
8. Subclass Placodontomorpha (euryapsid)
9. Subclass Ichthyopterygia (euryapsid)
10. Subclass Synapsida (synapsid)

The Labyrinthodontia of the Carboniferous, not the living amphibians, are considered as ancestors of reptiles. Small Anthracosauria (*Gephyrostegus, Brukterpeton*, figure 94) are quoted as the ancestral "model." However, these were already preceded by reptiles.

FIGURE 97. Body plan of the reptiles.
a, b Development of the embryonic membranes of Sauropsida (reptiles and birds); a, allantois, af, amniotic fold, ah, amniotic cavity, ch, chorion (serosa), ds, yolk sac, e, embryo. **c, d** Heart and aortic arches of Amphibia (**c**) and Reptilia (**d**); Ao, aorta, Aow, aortic stem, K, ventricle, Ka, carotids, La, pulmonary arteries, Luk, lung capillaries, V, atrium, 3, 4, 6, 3d, 4th, and 6th aortic arches. **e** Cross section of reptilian skin; e, epidermis, me, melanophore, o, osteoderm (dermal bone). **f** Longitudinal section of reptilian brain; Bo, bulbus olfactorius, H, metencephalon, M, mesencephalon, N, myelencephalon, V, prosencephalon, Z, diencephalon **g** Jacobson's organ. **h, i** Atlas-axis complex of a primitive reptile (**h**) and a mammal (**i**); c^1, c^2, centrums of the first two vertebrae, i^1, i^2, intercentrums, n^1, n^2, neural arches, pr proatlas. **k–m** Tooth attachment in reptiles, **k** Pleurodont, **l** Acrodont, **m** Thecodont. **n** Carpus, **o** Tarsus of a primitive reptile. Further abbreviations as in figure 87 (after *Bellairs, Kühn, Kuhn-Schnyder, Rietschel, Romer*).

TABLE 21.   Comparison of Amphibian, Reptilian, and Mammalian
Characteristics

Characteristic	Amphibians	Reptiles	Mammals
Body temperature	Variable	Variable	Constant
Breathing	Gills, lungs, skin	Lungs	Lungs
Breathing principle	Pressure	Suction	Suction
Skin	Mostly naked, glandular	Scales, few glands	Hair, scales, many glands
Eggs	Small supply of yolk	Abundant yolk	Small supply of yolk
Development	With metamorphosis	Without metamorphosis	Without metamorphosis
Embryonic membranes	Absent	Present	Present
Heart ventricle	Incompletely divided	Incompletely divided	Divided
Auditory ossicles	1	1	3
Occipital condyli	2	Usually 1	2
Atlas-axis complex	−	+	+

Two groups of the oldest reptiles are known to have occurred as early as the Upper Carboniferous: Protothyromorpha and Pelycosauria. The transition from amphibian to reptile apparently occurred without any basic changes in the skeletal structure, so that the only positive criterion to distinguish between the two is the kind of reproduction: Amphibia are Anamnia, Reptilia are Amniota.

1. Subclass Anapsida

Three orders are included in the Anapsida: Protothyromorpha, Procolophonia, and Pareiasauria. Temporal fenestrae are absent. The numerous specialized modifications of turtles (Testudines) justify their separation from Anapsida as a separate subclass, Chelonomorpha. Order Protothyromorpha assumes a key position because it contained ancestors of most higher reptiles.

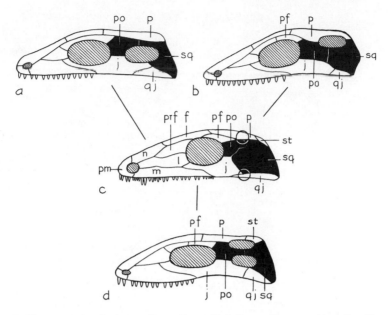

FIGURE 98. Temporal region of reptiles. **a** Synapsid, **b** Euryapsid, **c** anapsid, **d** diapsid. Abbreviations as in figure **88** (after *Kuhn-Schnyder*).

Order Protothyromorpha (figure 99)

Primitive, mostly small, insectivorous Anapsida (Carboniferous–Permian), with a high, narrow skull. Of the two families (**Romeriidae, Captorhinidae**), only Romeriidae deserve a designation as "original reptiles." Their oldest known representative is *Hylonomus* (Upper Carboniferous). A number of isolated finds from the Permian, with one temporal fenestra, have been considered related to the Protothyromorpha: *Eunotosaurus, Mesosaurus, Araeoscelis, Bolosaurus, Millerosaurus*.

Order Procolophonia (figure 100, a–d)

Small-bodied, lizardlike Anapsida (Permian–Triassic) with short skulls exhibiting otic notches. Differentiated dentition. During the Triassic, Procolophonia were distributed worldwide, including the Antarctic.

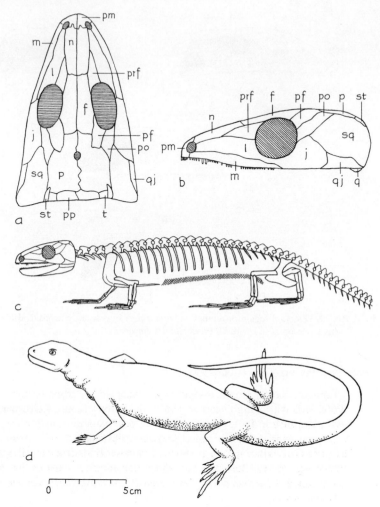

FIGURE 99. Protothyromorpha: *Hylonomus* (Upper Carboniferous). **a, b** Dorsal (**a**) and lateral (**b**) views of the skull; **c** Restoration of skeleton, length without tail approx. 40 cm; **d** Appearance of a live animal. Abbreviations as in figure **88** (after *Carroll*).

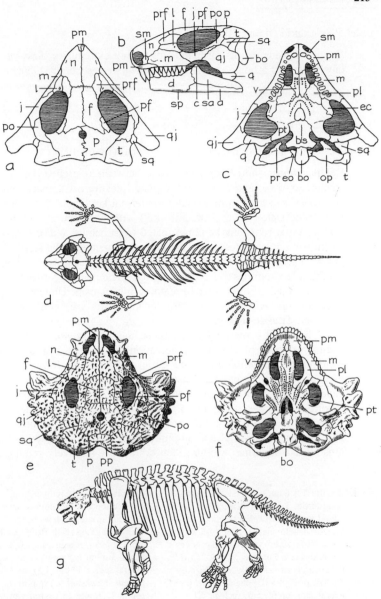

FIGURE 100. a–d Procolophonia: *Procolophon* (Lower Triassic). **a–c** Dorsal (**a**), lateral (**b**), and ventral (**c**) views of the skull, **d** Restoration of the skeleton, ≈30 cm. **e–g** Pareiasauria: *Scutosaurus* (Permian). **e, f** Skull in dorsal (**e**) and ventral (**f**) views, **g** Restoration of the skeleton, about 3 m long. Abbreviations as in figure **88** (after *Broili* and *Schröder, Efremov, Gregory*).

Order Pareiasauria (figure 100, e–g)

Large Anapsida of the Permian, with strong extremities. Several species possessed a bony armor. Pareiasauria were probably giant, swamp-dwelling herbivores.

2. Subclass Chelonomorpha

Order Testudines (Testudinata, Chelonia, turtles) (figure 101)

Reptiles whose short body is enclosed between an arched dorsal shield (carapace) and a flat ventral plate (plastron). Carapace and plastron are formed from an outer epidermal keratinous layer over an inner bone plate (theca). Akinetic skull without temporal fenestrae. The temporal roof may be emarginated from behind and/or below. There is a tendency to reduce the dermal elements of the skull. Teeth absent, except in the oldest forms. A horny bill developed in place of dentition. Extremities are pentadactylous. The phalangeal formula is mostly 2 3 3 3 3. Specialized breathing mechanism. Unpaired copulatory organ. Oviparous. Ancestors of the turtles are unknown. Distribution: Triassic–Recent.

The oldest known Triassic representatives (*Proganochelys*) already exhibit typical turtle organization. However, head and tail were not yet retractible under the carapace. Since the Jurassic, two evolutionary lines have been distinguished, based on the functional capabilities of the neck. **Cryptodira** retract their heads by a vertical flexure of the neck, forming an S-curve. **Pleurodira** bring their heads under the shield sideways. Cryptodira are a prosperous group, encompassing terrestrial and amphibic representatives in addition to those that are marine. Modern Pleurodira are a relatively small group

FIGURE 101. Testudines. **a** *Emys* (Recent, pond turtle, Testudinoidea), skull in lateral view. **b** *Archelon* (Upper Cretaceous, Chelonoidea), lateral view of the skull, ≈70 cm. **c, d** Dermal bones of the theca, **c** Carapace; mn I, mn II, metaneurals, n I–n VIII, neurals, nu, nuchal, pl I–pl VIII, pleurals, pn, preneural, ppl, propleural, pr I–pr XI, peripleurals, py, pygal, **d** Plastron; ent, entoplastron, ept, epiplastron, hyo, hyoplastron, hypo, hypoplastron, mpl, mesoplastron, xi, xiphiplastron. **e** Retraction of head in turtles, above Cryptodira, lateral view, below Pleurodira, dorsal view. **f–h** *Proganochelys* (Triassic, Proganochelidia), **f, g** Lateral (**f**) and ventral (**g**) views of the skull, **h** Carapace, ≈60 cm. **i** *Eretmochelys* (Recent, hawksbill, Cryptodira), skeleton and carapax in ventral view; fon fontanelle. Other abbreviations as in figures **87, 88** (after *Bellairs, Bojanus, Jaekel, Wieland, Williams, Zangerl*).

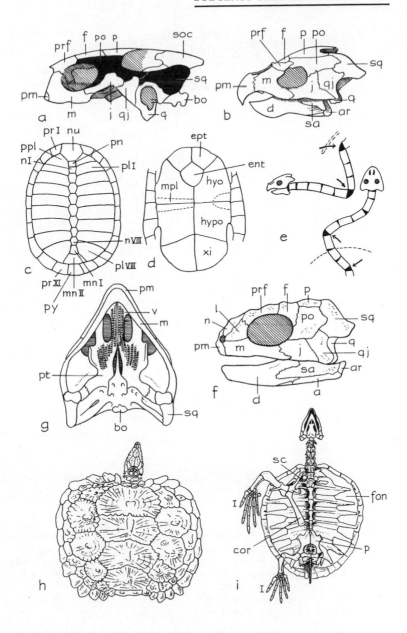

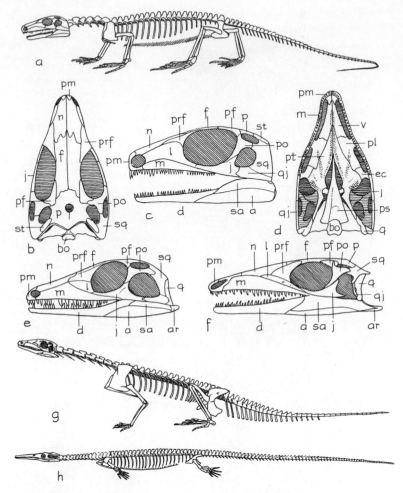

FIGURE 102. Eosuchia. a–d *Petrolacosaurus* (Upper Carboniferous, N. America), the oldest known Diapsid, **a** Restoration of skeleton, ≈60 cm, **b–d** Skull, dorsal view (**b**), lateral view (**c**), ventral view (**d**). **e** *Youngina* (Upper Permian, S. Africa), skull in lateral view. **f, g** *Prolacerta* (Lower Triassic, S. Africa, Antarctic), **f** Lateral view of the skull, **g** Restoration of the skeleton, ≈60 cm. **h** *Askeptosaurus* (Middle Triassic, Europe), restoration of skeleton, ≈2.5 m long. Abbreviations as in figure **88** (after *Carroll, Gow, Kuhn-Schnyder, Peabody, Reisz*).

of freshwater forms; they were widely distributed during the Cretaceous. Turtles remain phylogenetically conservative; the tertiary representatives almost entirely correspond to the modern ones.

3. Subclass Eosuchiamorpha

Order Eosuchia (figures 102, 103, a–c)

The oldest known Diapsida belong to this order. They emerged in Upper Carboniferous (*Petrolacosaurus*, N. America). A typical representative is *Youngina* (Upper Permian, S. Africa). In addition to these primitive, lizardlike representatives, specialized forms developed during the Triassic: aquatic tangasaurids, marine thalattosaurids, tanystropheids with extremely long necks (*Tanystropheus*), and even forms capable of flying (gliding; *Weigeltisaurus*, Permian, Europe). The lower zygomatic arch was reduced independently in various lines. Lepidosauria do derive from the Eosuchia; it is presumed that Archosauria, Pterosauria, and Sauropterygia also derive from Eosuchia.

4. Subclass Lepidosauria

Original Lepidosauria had a diapsid skull; subsequently one or both zygomatic arches may have disappeared. The monimostylic quadrate (firmly attached to the squamosal) would then become streptostylic (movably connected by an articulation). The palate is kinetic. The orders Rhynchocephalia (Triassic–Recent) and Squamata (Triassic–Recent) belong to this subclass.

Order Rhynchocephalia (figure 103, d–i)

Conservative Lepidosauria with a diapsid skull and specialized dentition. They encompass two suborders. **Sphenodontia** are small lizards that emerged during the Triassic. Never numerous, they are currently represented by a single species, the tuatara (*Sphenodon*). **Rhynchosauria** were highly specialized herbivores that had a worldwide distribution during the Middle Triassic and became extinct toward the end of the Triassic. They have sometimes been placed in a separate order.

220

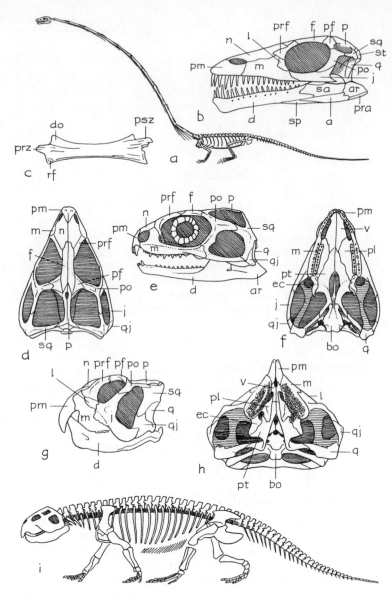

FIGURE 103. **a–c** Eosuchia: *Tanystropheus* (Middle Triassic). **a** Restoration of skeleton, over 6 m long, **b** Skull of an adult, lateral view, **c** Cervical vertebra; do, spinous process, prz, prezygapophysis, psz, postzygapophysis, rf, attachment site of a cervical rib. **d–f** Rhynchocephalia: *Sphenodon punc-*tatus (Recent, tuatara). Skull in dorsal (**d**), lateral (**e**), and ventral (**f**) views. **g–i** Rhynchosauria: *Scaphonyx* (Middle Triassic, S. Africa). Lateral (**g**) and ventral (**h**) views of the skull. **i** Restoration of skeleton, ≈1 m long. Abbreviations as in figure **88** (after *v. Huene, Romer, Wild*).

Order Squamata (figures 104, 105)

Lepidosauria with a modified diapsid skull. Extremities and girdles may be absent. The Jacobson's organ is well developed. Males of the living representatives have a paired copulatory organ, which distinguishes them from the living Archosauria. Oviparous or ovoviviparous. The most important phylogenic achievement of Squamata is a highly kinetic skull in which the anterior part can be raised. The lower jaw may also become kinetic (streptognathy). Squamata are divided into three suborders.

1. Suborder Eolacertilia (figure 104, c, d)

The original representatives are Palinguanidae (Permian–Triassic). They are descendants of Eosuchia. Gliding forms are known from the Upper Triassic (*Kuehneosaurus,* Europe; *Icarosaurus,* N. America).

2. Suborder Lacertilia (lizards) (figure 105, a–e)

Modern lizards can be traced back only as far as the Upper Jurassic. The following is a listing of the infraorders:

1. Infraorder Gekkota (Upper Jurassic–Recent)
2. Infraorder Iguania (Upper Jurassic–Recent)
3. Infraorder Scincomorpha (Upper Jurassic–Recent)
4. Infraorder Anguimorpha (Upper Cretaceous–Recent)
5. Infraorder Platynota (Upper Jurassic–Recent)

Of interest to paleontology are first of all the Platynota, middle-sized to very large predatory lizards with a highly kinetic skull. Three families turned to a marine way of life. Most impressive are the very large **Mosasauridae,** long-stretched, scaly lizards of the Cretaceous, whose limbs were transformed into paddle-shaped organs.

3. Suborder Ophidia (snakes) (figure 105, f–h)

Ophidia, or Serpentes, have an elongated, limbless body. Components of the skull can move easily against each other. Vertebrae with a zygosphen-zygantrum joint. Occasionally there are rudiments of the pelvic girdle. Several groups gained the capability of killing their prey by strangulation or poison. The question of snake origins is still open. It is believed that there are ties to the Platynota. The fossil record is scanty. The oldest remains are known from the Cretaceous.

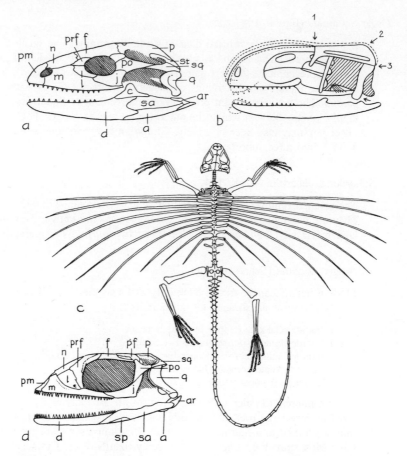

FIGURE 104. Squamata. **a** Diagram of a squamate skull, lateral view. **b** Kinetics of lizard skull (schematically); occipital segment cross-hatched, 1, anterior point of flexing (mesokinetic), 2, posterior point (metakinetic), 3, articulation between braincase and quadrate. **c, d** *Kuehneosaurus* (Upper Triassic), **c** Restoration of skeleton, approx. life-size, **d** Skull in lateral view. Abbreviations as in figure **88** (after *Goodrich, Robinson, Starck*).

5. Subclass Archosauria (ruling reptiles)

A large group of active, diapsid reptiles of medium to very large dimensions, with several bipedal lines. The skull is akinetic, mostly with an antorbital opening. Bipedalism led to profound modifications of girdles and limbs. Distribution: Permian–Recent.

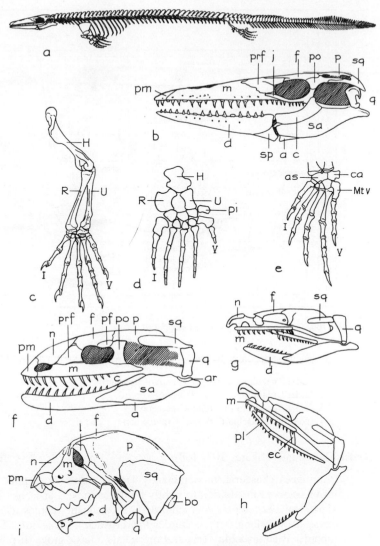

FIGURE 105. Squamata. **a, b** *Clidastes* (Upper Cretaceous, marine, Mosasauridae). **a** Restoration of the skeleton, ≈5 m long, **b** Lateral view of the skull. **c** *Varanus* (Recent, Platynota), forelimb. **d** *Clidastes*, front "paddle"; pi, pisiform. **e** *Varanus*, bones of the foot; as, astragalus, ca, calcaneus, MtV, hamatum. **f** Schematic drawing of a snake skull, in lateral view. **g, h** Snake skull kinetics. **i** *Agamodon* (Recent, Caecilia, Amphisbaenidae). Lateral view of the skull. Abbreviations as in figures **87, 88** (after *Bellairs, Goodrich, Lehman, Starck, Versluys, Williston*).

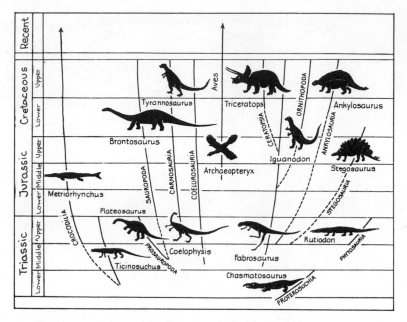

FIGURE 106. Phylogenetic tree of Archosauria (after *Krebs*).

Archosauria are divided into the following orders (figure 106):

1. Order Thecodontia (Upper Permian–Triassic)
2. Order Crocodilia (Upper Triassic–Recent)
3. Order Saurischia (Triassic–Cretaceous)
4. Order Ornithischia (Triassic–Cretaceous)

Order Thecodontia (figures 107, 108)

Primitive Archosauria, predominantly Triassic. Heavily armored forms occurred in addition to lightly built quadrupeds and bipeds. Thecodontia encompass four suborders. Semiaquatic reptiles (*Chasmatosaurus*) belong to the suborder **Proterosuchia** (Triassic). The suborder **Pseudosuchia** (Triassic) represents a large group of predatory thecodonts of extraordinary plasticity. A generalized form is represented by *Euparkeria*. Later the Pseudosuchia evolved along two lines. Rauisuchids were fast quadrupeds whose tracks are strikingly similar to those of *Chirotherium*. Ornithosuchidae perfected bipedalism. The suborder **Aetosauria** (Triassic) contains large, armored forms. **Phytosauria** (Triassic) are a closely knit group of large, car-

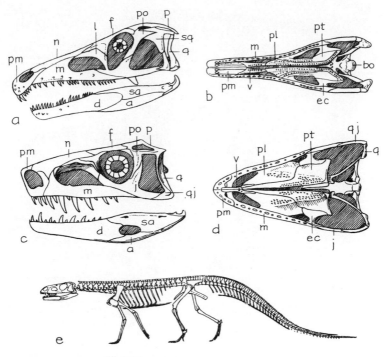

FIGURE 107. Thecodontia. **a, b** *Chasmatosaurus* (Lower Triassic, Proterosuchia), 45 cm. **c, d** *Euparkeria* (Lower Triassic, Pseudosuchia), lateral (**c**), ventral (**d**) views of the skull, ≈9 cm. **e** *Ticinosuchus* (Middle Triassic, Pseudosuchia), restoration of skeleton, ≈2.5 m. Abbreviations as in figure **88** (after *Cruickshane, Ewer, Krebs*).

nivorous, long-snouted reptiles whose appearance resembled that of the crocodiles.

Order Crocodilia (figure 109)

Semiaquatic or aquatic quadrupedal Archosauria of medium to considerable size. The antorbital fenestra is reduced or absent. Premaxilla, maxilla, palatines, and pterygoids may join in the middle and form a secondary palate, thus separating the food passage from the breathing passage. Rows of dermal bones may appear on the back and belly. In the living forms the heart ventricle is almost completely

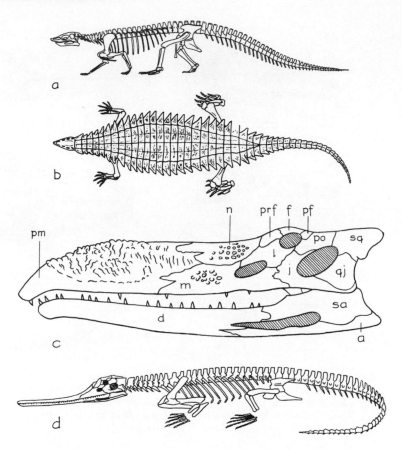

FIGURE 108. a, b Aetosauria: a *Stagonolepis* (Triassic), restoration of skeleton, ≈3 m, b *Typothorax* (Triassic), restoration of skeleton, showing the dermal-bone armor, ≈3 m long. c, d Phytosauria: *Nicrosaurus* (*Phytosaurus* s.l. in part) (Upper Triassic), lateral view of the skull, ≈0.65 m, d *Mystriosuchus* (Triassic), restoration of skeleton, ≈4 m long. Abbreviations as in figure 88 (after *Fraas, McGregor, Sawin, Walker*).

FIGURE 109. Crocodilia. a Diagram of crocodilian skull, lateral view. b, c *Proterosuchus* (Upper Triassic–Lower Jurassic, Proterosuchia), b Skull in lateral view, c Restoration of skeleton, ≈0.8 m. d–g *Geosaurus* (Upper Jurassic, Metriorhynchidae), d Lateral view of the skull, e Manus, f Pes, g Appearance of a live animal, ≈2 m long. h *Baurusuchus* (Upper Cretaceous, Sebecosuchia), lateral view of the skull, ≈0.4 m. i, k *Alligator* (Recent, Eusuchia), i Manus, proximal carpalia extended, k Pes. l *Gavialis* (Recent, Eusuchia), ventral view of the skull, partial view of the secondary palate, choanae (ch) located posteriorly. Abbreviations as in figures 87, 88 (after *Colbert* and *Mook, Fraas, Goodrich, Kälin, Knüsel, Price*).

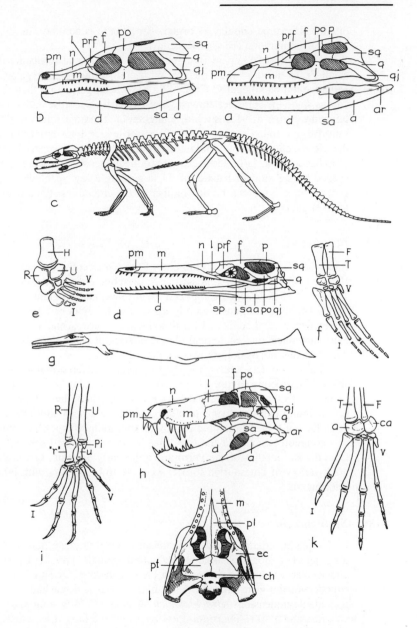

divided. Unpaired copulatory organ. Brood care is common in most forms. Distribution: Triassic–Recent.

Crocodilia encompass five suborders. Transitional forms are the small **Protosuchia** of the Triassic. While remaining relatively conservative morphologically, Crocodilia have been perfecting their adaptations since the Jurassic. Jurassic and Cretaceous representatives are lumped together as **Mesosuchia**. Long-snouted marine forms (Teleosauridae) are found among them. Two sterile side lines branched from the Mesosuchia: pelagic **Thalattosuchia,** and a small terrestrial group of **Sebecosuchia,** large terrestrial crocodiles. Mesosuchia were succeeded by the modern **Eusuchia**. Disregarding a few specialized genera, three types can be distinguished: (1) crocodiles, (2) alligators, and (3) gavials.

DINOSAURIA

From the Middle Triassic to the end of the Cretaceous, the land was ruled by two orders of Archosauria: **Saurischia** and **Ornithischia**. The secret of their success was fast locomotion and efficient dentition. The dinosaurs are currently a focus of lively debate concerning whether they were cold- or warm-blooded. In view of the variety of their anatomical and ecological characteristics, it can be conjectured that their thermophysiology was variable as well.

The two orders Saurischia and Ornitischia differed mainly in the shape of the pelvis (figure 110, a, b). The saurischian pelvis is triradiate, with the pubis directed forward. Ornithischia have a posteriorly pointing pubis that runs parallel to the ischium; in addition, they often have a forward directed processus (tetraradiate type). Currently it is common to consider the two lines as having evolved separately from diapsid reptiles of the Permian or the early Triassic. Given the great variety of dinosaurian forms, we have to be content with little information.

Order Saurischia (figures 110, 111)

Fossil quadrupedal or bipedal Archosauria with triradiate pelvis and long tail. The foot is birdlike. Distribution: Middle Triassic–Upper Cretaceous. Evolution followed two lines. Quadrupedal **Sauropodomorpha** were predominantly herbivorous and reached formidable dimensions. *Brachiosaurus* was about 24 m long and weighed about 50 t. **Theropoda** evolved into the largest terrestrial predators (*Tyrannosaurus*).

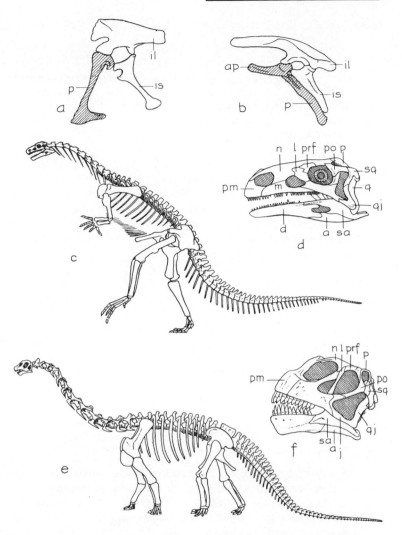

FIGURE 110. a, b Dinosauria: Pelvis, pubis cross-hatched. **a** Saurischia, **b** Ornithischia. **c–f** Saurischia: **c, d** *Plateosaurus* (Upper Triassic, Prosauropoda), **c** Restoration of skeleton, up to 7 m long, **d** Skull, lateral view. **e, f** *Camarosaurus* (Upper Jurassic–Lower Cretaceous, Sauropoda), **e** Restoration of skeleton, ≈6 m, **f** Skull, lateral view. Abbreviations as in figures **87, 88** (after *Gilmore, v. Huene*).

Order Ornithischia (figures 112, 113)

Fossil qudrupedal or bipedal, mainly herbivorous Archosauria with tetraradiate pelvis. The anterior part of the jaw is often toothless, covered by horny plates. Dermal bone armor may occur. Distribution: Upper Triassic–Upper Cretaceous. Ornithischia contain a variety of bizarre types. **Ornithopoda** are relatively primitive, bipedal forms. Famous is the *Iguanodon;* the remains of a herd of about thirty individuals of this genus were found in a Belgian coal mine. A number of genera carried grotesque skull outgrowths, ridges, and tubular structures that enclosed a long nasal passage. Large triangular bony plates on the back of the **Stegosauria** are considered to have had a thermoregulatory function. Quadrupedal **Ancylosauria** were heavily armored in large and small bony plates. The last of the Ornithischia were Upper Cretaceous **Ceratopsia**. Typical representatives had a large bony shield over the neck region. *Triceratops* carried a median horn in addition to a pair of large horns.

All dinosaurs, with the exception of Prosauropoda and Stegosauria, lived to the end of the Mesozoic. New types appeared even during the Upper Cretaceous. At the beginning of the Tertiary, all dinosaurs vanished. Why they, as well as numerous other animal groups of the Upper Cretaceous, became extinct is still an unsolved problem. It is assumed that the catastrophe was a consequence of the combined effects of geological, climatic, and ecological changes. But who would not be compelled to think about the ascent, rule, and fall of ancient and modern peoples?

6. Subclass Pterosauromorpha

Order Pterosauria (flying reptiles, pterosaurs) (figures 114, 115)

A well-defined group of very small to very large fossil reptiles that acquired an ability to fly. They are classified as a separate subclass because, unlike the Archosauria, they cannot be derived from the thecodonts. The lightly built skull is diapsid and akinetic, with a large antorbital fenestra. Dentition is limited to the jaw edges or replaced by a horny bill. The tail is long or short; the ossified sternum is large. Wings are supported by elongated phalanges of the IVth ray of the hand. The body is covered with hairy scales. Bones are hollow and extraordinarily light. Distribution: Upper Triassic–Upper Cretaceous.

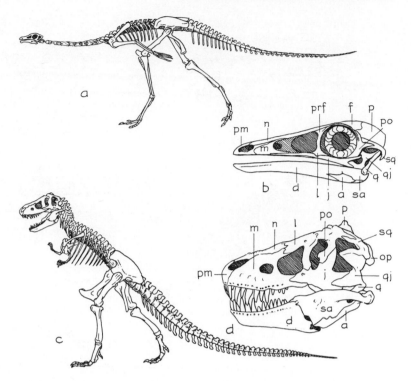

FIGURE 111. Saurischia. **a, b** *Ornithomimus* (Upper Cretaceous, Coelurosauria), **a** Restoration of skeleton, ≈3 m long, **b** Lateral view of the skull. **c, d** *Tyrannosaurus* (Upper Cretaceous, Car- nosauria), **c** Restoration of skeleton, over 12 m long, live weight 8 t, **d** Skull, lateral view. Abbreviations as in figure **88** (after *Gregory, Osborn*).

Two suborders are distinguished by length of the tail:

1. Suborder Rhamphorhynchoidea (long-tailed, Upper Triassic–Upper Jurassic): *Eudimorphodon, Rhamphorhynchus*.
2. Suborder Pterodactyloidea (short-tailed, Upper Jurassic–Cretaceous): *Pterodactylus, Pteranodon*.

The very high metabolic demands of flying require a high body temperature. The assumption of homoiothermy is supported by the insulating cover of hairy scales. Many pterosaurs were relatively small and active fliers. As they grew in size, they adopted thermal soaring. The best-known representative is *Pteranodon*, with a wing-

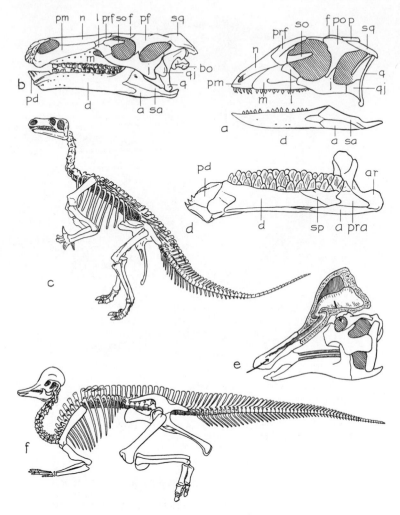

FIGURE 112. Ornithischia. **a** *Fabrosaurus* (Upper Triassic, Ornithopoda), lateral view of the skull, so, supraorbital, **b** *Camptosaurus* (Upper Jurassic–Lower Cretaceous, Ornithopoda), skull, lateral view. **c, d** *Iguanodon* (Upper Jurassic–Lower Cretaceous, Ornithopoda), **c** Skeleton, 4–5 m high, **d** Lower jaw, viewed from the inside. **e, f** *Corythosaurus* (Upper Cretaceous, Ornithopoda), **e** Outline of the skull with a restoration of the nasal passage; pm and n of the left side removed, **f** Restoration of skeleton, ≈10 m long. Abbreviations as in figure **88** (after *Colbert, Dollo, Gilmore, Ostrom, Thulborn*).

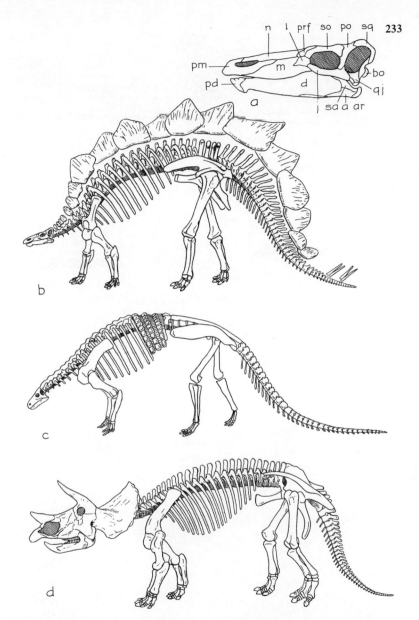

FIGURE 113. Ornithischia. **a, b** *Stegosaurus* (Upper Jurassic, Stegosauria). **a** Skull, lateral view, so, supraorbital. **b** Restoration of skeleton, up to 8 m long. **c** *Nodosaurus* (Upper Cretaceous, Ancylosauria), restoration of skeleton, ≈2.7 m long. **d** *Triceratops* (Upper Cretaceous, Ceratopsia), restoration of skeleton, 4–6 m long. Abbreviations as in figure **88** (after *Gilmore, Lull, Marsh*).

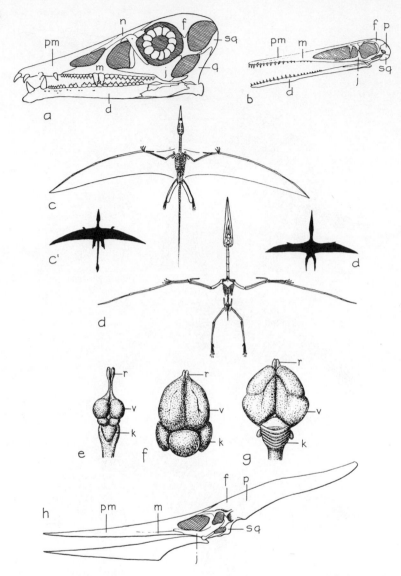

FIGURE 114. Pterosauria. **a** *Eudimorphodon* (Upper Triassic, Rhamphorhynchoidea), lateral view of the skull, ≈10 cm long. **b** *Pterodactylus* (Upper Jurassic, Pterodactyloidea), lateral view of the skull, 10 cm long. **c** *Rhamphorynchus* (Upper Jurassic), restoration of skeleton, with a flying silhouette. **d** *Pterodactylus* (Upper Jurassic), restoration of skeleton, with a flying silhouette. **e–g** Comparison of brains of a reptile (**e**), a flying saurian (**f**), and a bird (**g**). Brain of flying saurians resembles that of birds. **h** *Pteranodon* (Upper Cretaceous, Pterodactyloidea), lateral view of the skull, ≈1.7 m long. Abbreviations as in figure **88** (after *Wellnhofer, Wild*).

FIGURE 115. Wing structures. **a** Flying saurian; I thumb; **b** Bird; **c** Bat (after *Simpson*).

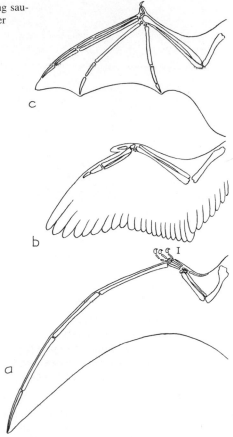

spread of 7–10 m. Even larger was *Quetzalcoatlus* (Upper Cretaceous) with a span of 11–12 m. As the pterosaurs are documented from the Upper Triassic as already having two distinct, highly developed families, the acquisition of the ability to fly must have occurred still earlier.

7. Subclass Sauropterygomorpha

Order Sauropterygia (figure 116)

Fossil, small to very large marine reptiles with a long neck and an akinetic skull with upper temporal fenestrae. The pentadactylous

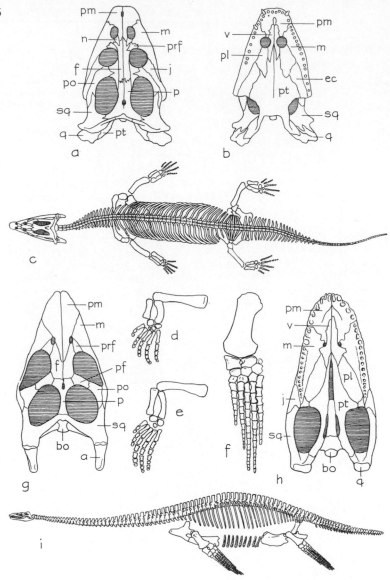

FIGURE 116. Sauropterygia. **a, b** *Simosaurus* (Middle Triassic, Nothosauria), skull, ≈35 cm. Dorsal **(a)** and lateral **(b)** views. **c** *Paranothosaurus* (Middle Triassic, Nothosauria), restoration of skeleton, ≈4 m long. **d–f** Hindlimbs. **d** *Lariosaurus* (Middle Triassic, Nothosauria), terrestrial phalangeal formula; **e** *Ceresiosaurus* (Middle Triassic, Nothosauria), beginning hyperphalangy, phalangeal formula 2 3 5 6 6; **f** *Muraenosaurus* (Upper Jurassic, Plesiosauria), very distinct hyperphalangy. **g–i** *Muraenosaurus*. **g, h** Skull, ≈35 cm, dorsal **(g)** and ventral **(h)** views; **i** Restoration of skeleton, ≈8 m. Abbreviations as in figure **88** (after *Andrews, Kuhn-Schnyder, Peyer*).

limbs were used as paddles. Phalangeal formula originally 2 3 4 5 3 (4), later was common hyperphalangy. Distribution: Upper Permian–Cretaceous.

Sauropterygian ancestors are sought among the Eosuchia. The structure of the temporal region suggests that they had passed through a diapsid stage. Two orders, **Nothosauria** (Upper Permian–Triassic) and **Plesiosauria** (Middle Triassic–Upper Cretaceous), are distinguished. Some of the up to 4-m-long Nothosauria show the beginnings of hyperphalangy. **Plesiosauroidea** were long-necked forms with a small skull. **Pliosauroidea** had a relatively short neck and a large skull. Both superfamilies, significantly hyperphalangeous, were adapted to life on the high seas. Plesiosauria cannot be derived from Nothosauria, yet they must have had common ancestors.

8. Subclass Placodontomorpha

Order Placodontia (figure 117)

Fossil aquatic reptiles of the Triassic, with specialized crushing teeth. The skull is akinetic, with one temporal fenestra. Dentition is markedly heterodont; in advanced forms it is reduced. The phalangeal formula is unknown. There is a tendency to form a dermal armor. The origin of Placodontia remains unknown. They are not in any way related to the Sauropterygia. The following three suborders are distinguished by the type of dentition:

1. Suborder Placodontoidea

Paraplacodus, Placodus, with dermal ossifications over the thoracic vertebrae. Marine. Distribution: Middle Triassic.

2. Suborder Cyamodontoidea

Cyamodus, Psephoderma, Placochelys. Heavily armored, some with a horny beak. Marine. Distribution: Middle–Upper Triassic.

3. Suborder Henodontoidea

Henodus, heavily armored, with toothless, wide snout with horny plates. Freshwater. Distribution: Upper Triassic.

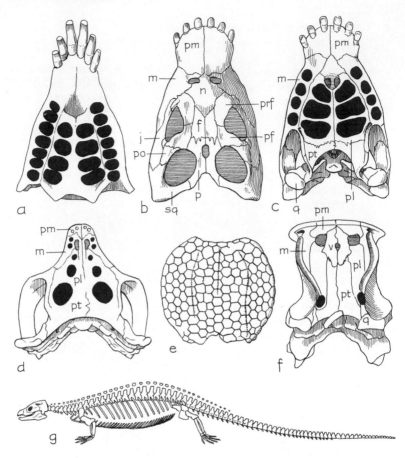

FIGURE 117. Placodontia. **a** *Paraplacodus* (Middle Triassic, Placodontoidea), upper dentition. **b, c** *Placodus* (Triassic, Placodontoidea), skull, dorsal (**b**), and ventral (**c**) views. **d** *Cyamodus* (Middle Triassic, Cyamodontoidea), ventral view of the skull. **e** *Psephoderma* (Triassic, Cyamodontoidea), dermal armor. **f** *Henodus* (Upper Triassic, Henodontoidea), ventral view of the skull. **g** *Placodus* (Triassic), restoration of skeleton, up to 3 m long. Abbreviations as in figure **88** (after *Broili, v. Huene, Kuhn-Schnyder, Peyer, Westphal*).

9. Subclass Ichthyopterygia

Order Ichthyosauria (ichthyosaurs) (figure 118)

Fossil marine reptiles, with a fusiform body, fin-shaped extremities, one dorsal fin, and one vertical tail fin. Akinetic skull with upper temporal fenestrae. Long vertebral column. Stylopodium and zeugopodium shortened. The number of rays of the autopodium may be increased (hyperdactyly) or reduced. Number of phalanges is increased (hyperphalangy). Ovoviviparous. Distribution: Triassic–Cretaceous.

The origin of Ichthyosauria is unknown. Small, broad-finned **Mixosauridae** of the Triassic are not yet quite as specialized as the other representatives. Very large, narrow-finned **Shastasauridae** (up to 10 m) were their contemporaries. Broad-finned **Ichthyosauridae** are known from the Jurassic to the Cretaceous periods. The preferred prey of many ichthyosaurs were cuttlefish. Ideal adaptation to marine life made it impossible for ichthyosaurs to lay eggs on dry land. Their few embryos developed inside the mother's body and were born alive. Ichthyosaurs became more and more rare during the Cretaceous and vanished with the end of this period.

10. Subclass Synapsida (Theromorpha)

Synapsida bridge the gap between reptiles and mammals. Forms with akinetic, synapsidlike skulls are already represented in the oldest reptile faunas. They have mostly heterodont dentition. The pectoral girdle consists of scapula, procoracoid and metacoracoid, clavicula and interclavicula, and originally a rudimentary cleithrum. All synapsids were quadrupedal, usually with slim extremities.

Distribution: Upper Carboniferous, Lower Jurassic.

There are two orders of Synapsida:

Order Pelycosauria (Upper Carboniferous–Lower Permian)
Order Therapsida (Middle Permian–Lower Jurassic)

Pelycosauria are known mainly from N. America and Europe; Therapsida mainly from Africa, Russia, and S. America. Synapsida were a dominant group during the Paleozoic era. During the Triassic, they became overshadowed by the rapidly developing Archosauria.

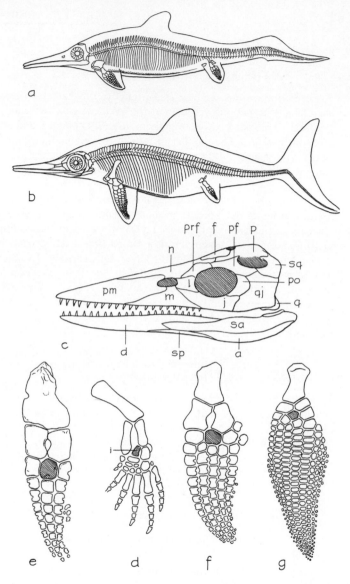

FIGURE 118. Ichthyosauria. **a** *Mixosaurus* (Triassic, Mixosauridae), restoration of skeleton, up to 1.5 m long. **b** *Ichthyosaurus* (Jurassic, Ichthyosauridae), restoration of skeleton, up to 10 m or longer; **c** Diagram of an ichthyosaurian skull, lateral view. **d–g** Forelimbs. **d** original reptilian; **e** *Merriamia* (Triassic, Shastasauridae); **f** Mixosaurus; **g** *Eurypterygius* (Jurassic, Ichthyosauridae). Abbreviations as in figure **88** (after *Goodrich, Jaekel, Kuhn-Schnyder, Stromer v. Reichenbach*).

Order Pelycosauria (figure 119)

This original group of the Synapsida appeared during the Upper Carboniferous and is osteologically closely related to its contemporary group of Romeriidae (Protothyromorpha). Their initially small representatives soon became larger and evolved in many different directions. There were long-snouted fisheaters (**Ophiacodontia**), dangerous carnivores (**Sphenacodontoidea**), and short-mandibled herbivores (**Edaphosauria**). *Dimetrodon* (Permian, N. America, Sphenacodontidae), 2.5 m long, had unusually elongated spinous processes over which was stretched skin richly supplied with blood vessels. This "sail" had a thermoregulatory function. A similar sail was convergently developed by the *Edaphosaurus* (Upper Carboniferous–Lower Permian; Europe, N. America). Pelycosauria were succeeded by the Therapsida. However, true connecting links are so far unknown.

Order Therapsida (figures 120, 121)

Therapsida are extraordinarily polymorphic synapsids that flourished during the Upper Permian and Lower Triassic. Because of variable evolutionary trends, there is no single diagnostic characteristic that would be valid for all of them. Two main branches (suborders) evolved from primitive **Phtinosuchia** (Middle Permian, Europe).

1. Suborder Anomodontia

Ancient representatives of the herbivorous Anomodontia are **Dinocephalia,** which tended to grow into giant forms (*Ulemosaurus,* Middle Permian, Europe; *Moschops,* Middle Permian, S. Africa). Dinocephalia diverged early from **Dicynodontia,** which lived in large numbers in the Permian Africa, many of them in swamps. Their dentition was reduced and replaced by horny plates. *Dicynodon,* rat-sized to ox-sized, possessed a pair of upper fangs (often absent). Aquatic *Lystrosaurus* (Lower Triassic) was also found in the Antarctic. Late forms such as *Kannemeyeria* (Triassic, Africa, S. America) reached considerable size.

2. Suborder Theriodontia

Mostly carnivorous Therapsida, whose various branches independently evolved toward mammalian characteristics. Sometimes skull characteristics, sometimes postcranial skeletal characteristics

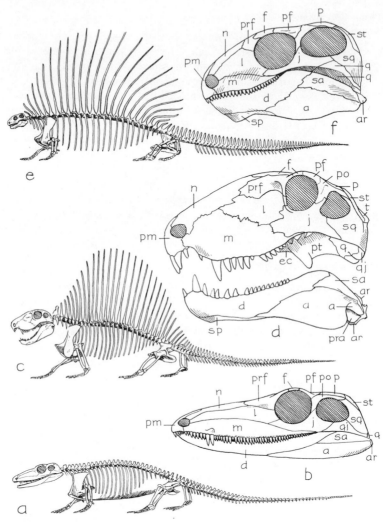

FIGURE 119. Pelycosauria. **a, b** *Varano-saurus* (Lower Permian, Ophiacodon-tia). **a** Restoration of skeleton; approx. 1.5 m long; **b** Lateral view of the skull. **c, d** *Dimetrodon* (Lower to Middle Permian, Sphenacodontoidea). **c** Resto-ration of skeleton; over 3 m long; **d** Lateral view of the skull. **e, f** *Edaphosaurus* (Upper Carbonife-rous–Lower Permian, Edaphosauria). **e** Restoration of skeleton; approx. 3 m long; **f** Lateral view of the skull. Abbreviations as in figure **88** (after *Romer*).

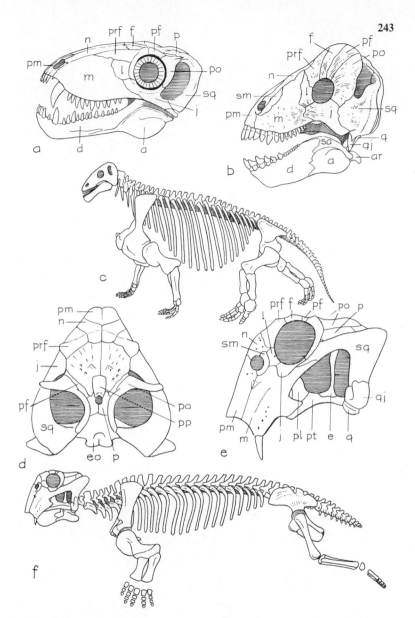

FIGURE 120. Therapside. **a** *Phtinosuchus* (Middle Permian, Phtinosuchia), lateral view of the skull, ≈20 cm long. **b** *Ulemosaurus* (Middle Permian, Dinocephalia), lateral view of the skull, ≈40 cm. **c** *Moschops* (Middle Permian, Dinocephalia), restoration of skeleton, up to 3 m long. **d–f** *Lystrosaurus* (Lower Triassic, Dicynodontia), skull in dorsal (**d**) and ventral (**e**) views, **f** Restoration of skeleton, ≈1.2 m. pp, preparietal. Other abbreviations as in figure **88** (after *Colbert, Efremov, Gregory, Orlov*).

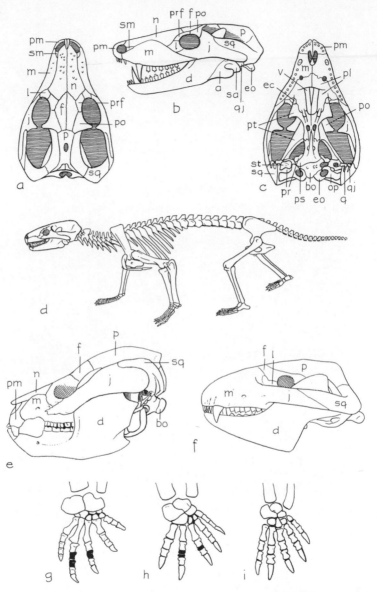

FIGURE 121. Therapsida. **a–d** *Thrinaxodon* (Lower Triassic, Cynodontia), dorsal **(a)**, lateral **(b)**, and ventral **(c)** views of the skull, **d** Restoration of skeleton, ≈0.5 m long. **e** *Bienotherium* (Upper Triassic, Tritylodontoidea), lateral view of the skull, ≈12 cm. **f** *Diarthrognathus* (Upper Triassic, Ictidosauria), lateral view of the skull, ≈4 cm long.

g–i Hind feet of Synapsida, reduction in the number of phalanges; **g** *Ophiacodon* (Lower Permian, Pelycosauria); **h** *Lycaenops* (Lower Permian, Therapsida); **i** *Whaitsia* (Lower Permian, Therapsida). Abbreviations as in figure **88** (after *Crompton, Hopson, Jepsen, Romer*).

would become mammallike. Examples of such trends are the increase in size of the temporal opening, reduction of the dermal skull components and the postorbital bar, and development of a secondary palate from parts of the premaxillae, maxillae, and palatines (separation of air and food passages). Other characteristics are reduction of quadratojugal and quadrate, enlargement of the dentary, differentiation of the dentition into incisors, canines, and molars, positioning of the extremities underneath the body, and reduction of the middle phalanges leading to a phalangeal formula 2 3 3 3 3.

Very large forms occurred among the primitive Theriodontia. *Jonkeria* (Middle Permian, S. Africa) reached a length of 4 m. Of special interest are small, lightly built **Cynodontia,** that had a secondary palate, double occipital condyle, large dentary, and a heterodont dentition (*Thrinaxodon,* Lower Triassic, S. Africa, Antarctic; *Cynognathus,* Lower to Middle Triassic, S. Africa). *Probainognathus* (Middle Triassic, S. America) had functional primary and secondary jaw articulations. Proof of tactile hairs in the progressive theriodonts indicates that these mammallike reptiles had hairy skin and more or less constant body temperature. The roots of mammals are sought among the progressive Cynodontia.

7. Class Aves (birds)

The close relationship of birds and reptiles was already known to THOMAS H. HUXLEY (1861), who united the two groups under the name Sauropsida. Nearly all characteristics separating birds from reptiles can be ascribed to their ability to fly. Flying puts heavy demands on every system of organs. Living birds have the following characteristics (figure 122):

1. Body covered with feathers.
2. Front extremities transformed into wings, rear extremities usually with four toes (I–IV). Tarsus and foot clad in horny scales.
3. Skeleton is fully ossified, most bones are pneumatized. Jaws either toothed or structured as a horny bill. Skull is derived from the diapsid type. Quadrate is streptostylic. One occipital condyle. Quadrato-articular joint. One auditory ossicle (columella = stapes). Neck very flexible and mobile. Many sacral vertebrae. Last caudal vertebrae fused (pygostyle). Clavicles mostly distally fused (furcula). Large sternum, generally with a keel. Pelvis is elongated both before and after the acetabulum. Pubis long and thin, running backward, parallel to the ischium. Carpus and skel-

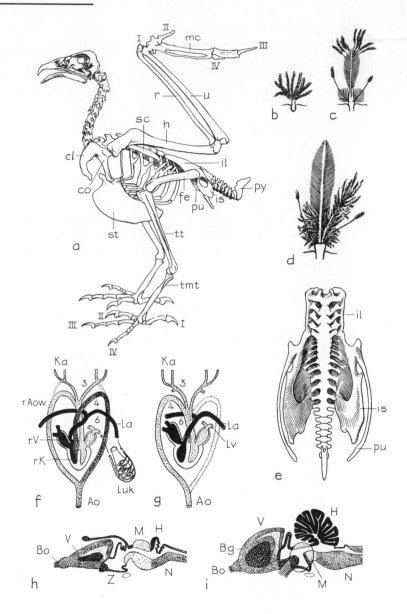

eton of the hand considerably reduced (with the exception of the IIId ray). The proximal tarsal row of the rear extremities is fused with the tibia (tibiotarsus), and the distal row is fused with metatarsals (tarsometatarsus). Fibula distally reduced.

4. Heart fully divided (two atria and two ventricles); three arterial stems, left aortic arch reduced.
5. Breathing with lungs connected to air sacs.
6. No urinary bladder, excreta semiliquid. Females have developed only the left ovary and left oviduct.
7. Cerebrum well developed, cerebellum very well developed. Twelve pairs of cranial nerves. Sight and hearing very efficient; sense of smell weak.
8. Body temperature constant (homoiothermy).
9. Fertilization internal. Oviparous, eggs rich in yolk. Embryonic membranes (amnion, chorion, allantois). Brood care (with few exceptions).

Distribution: Jurassic–Recent

Formation of feathers is a key avian characteristic. Feathers are important for thermal insulation as well as for flight. They are usually regarded as homologous with reptilian scales. The skull is prokinetic, that is, the upper half of the bill can be raised. All birds are adapted to specific types of flight (gliding, soaring, flapping, fluttering, or hovering). Loss of the ability to fly is connected with reduction of wings.

Birds are the most sharply delimited group of all vertebrates, yet they show an amazing variety of form. Dividing them into higher systematic categories is difficult. The fragility of their hard parts, as well as the type of habitat they choose, obviously does not favor fossil preservation.

FIGURE 122. Body plan of birds. **a** Skeleton of a vulture; cl, clavicle (furcula), co, coracoid, fe, femur, h, humerus, il, ilium, is, ischium, mc, metacarpal, pu, pubis, py, pygostyle, r, radius, sc, scapula, st, sternum, tmt, tarsometatarsus, tt, tibiotarsus, u, ulna.
b–d Feather development (chicken). **b** Down of a nestling, **c** Development of the vane of a contour feather; on the sides are two filoplumes. **d** Contour feather with an aftershaft and two filoplumes. **e** Pelvis (synsacrum) of a chicken, ventral view. **f, g** Heart and aortic arches of a reptile (**f**) and a bird (**g**); Ao, aorta, Aow, aortic stem, K, ventricle, Ka, carotid, La, pulmonary artery, Luk, lung capillaries, Lv, lung vein, V, atrium, 3, 4, and 6 are aortic arches. **h, i** Brain of a reptile (**h**) and a bird (**i**); Bg, basal ganglion, Bo, bulbus olfactorius, H, cerebellum, M, mesencephalon, N, myelencephalon, V, prosencephalon, Z, diencephalon (after *Kühn, Portmann*).

1. Subclass Archaeornithes

Order Archaeopterygiformes (figures 123, 124, a)

The oldest avian remains of the Upper Jurassic, seven finds of one kind (*Archaeopteryx*), all come from the Solnhofen lithographic limestones of the Northern Frankonian Jurassic in Bavaria. *Archaeopteryx* still had predominantly reptilian characteristics (table 22). However, it is classified as a bird because it already had feathers.

The origins of the *Archaeopteryx* are sought among the large pool of the Archosauria. Opinions differ as to the "Proavis," its hypothetical predecessor. According to one hypothesis, it was a primitive arboreal climber that acquired the ability to glide. Another view makes it a bipedal runner that knocked down living prey with its forelimbs. *Archaeopteryx* cannot be regarded as a direct ancestor of the Cretaceous birds.

2. Subclass Neornithes

Approximately 9,000 bird species are known today, as opposed to about 1,000 fossil species. About half of all species belong to Passeriformes (perching birds). Three superorders were established, based on the presence or absence of teeth and on the structure of the skull base:

1. Superorder †Odontognathae (toothed birds), Cretaceous
2. Superorder Palaeognathae, Cretaceous–Recent
3. Superorder Neognathae, Cretaceous–Recent

1. Superorder Odontognathae (toothed birds) (figure 124, e, f)

Few fossil species of the Cretaceous, with toothed jaws. Order **Hesperornithiformes** is comprised of highly specialized, flightless divers and swimmers with reduced forelimbs. To the order **Ichthyornithiformes** belong birds with well-developed wings. By the end of the Mesozoic, ancestors of many modern orders had already appeared.

2. Superorder Palaeognathae (figures 124, c, 125, b)

Mostly large, toothless, omnivorous birds, most often unable to fly. Palate palaeognathous (large vomers). Sternum without a keel. To this group belong ostriches, rheas, cassowaries, kiwis, and tinamous, as well as the extinct Madagascar ostriches and the moas. Their

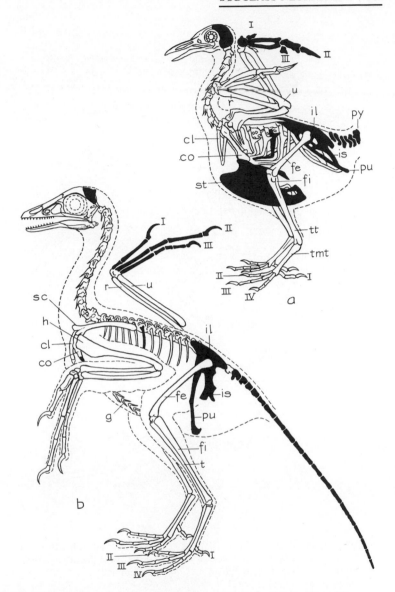

FIGURE 123. a Skeleton of a pigeon. **b** Skeleton of the *Archaeopteryx* (Jurassic); g, gastralia. Modified components in black. Abbreviations as in figure **122** (after *Heilmann*).

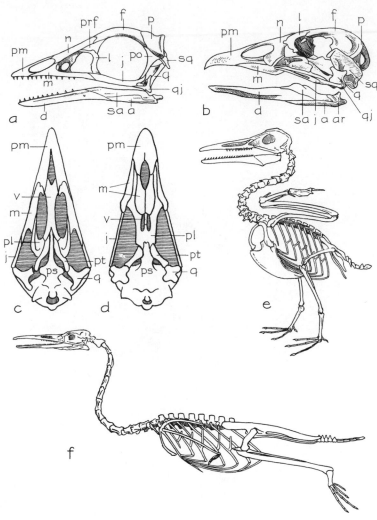

FIGURE 124. a *Archaeopteryx* (Jurassic), restoration of the skull, lateral view, ≈4.3 cm long. **b** *Anser* (Recent, goose), skull, young, lateral view. **c, d** The palate, **c** Palaeognathous, **d** Neognathous. **e** *Ichthyornis* (Upper Cretaceous, Ichthyornithiformes), restoration of skeleton, height ≈20 cm. **f** *Baptornis* (Upper Cretaceous, Hesperornithiformes), ≈0.7 m long. Abbreviations as in figure **88** (after *Heilmann, Marsh, Martin, Romer, Wellnhofer*).

TABLE 22.   Reptilian and Avian Characteristics of the Archaeopteryx

Reptilian	Avian
Simple brain, small cerebellum	True feathers
Toothed jaws	Furcula
Amphicoelous vertebrae, 2–3 sacral vertebrae	(connected clavicles)
Long tail, 22 vertebrae	
Ribs without processus uncinati	
Free metacarpals	
Long fibula	
Free metatarsals	
Sternum small and flat	
Gastralia	
Phalanges with claws	

ancestors were found as early as in the Cretaceous. **Aepyornithiformes** (Eocene–Subrecent) are known primarily from the Holocene of Madagascar. The largest species was about 3 m tall; its eggs had a volume of 6–10 l. **Dinornithiformes** (moas), from turkey-sized to more than 3-m-tall giants, are found in quaternary deposits of New Zealand. The common characteristics of these birds (Ratitae) are related to the loss of their ability to fly.

3. Superorder Neognathae (Carinata, modern birds) (figures 124, d, 125, a)

Toothless birds with a sternal keel, well-developed wings, and a neognathous palate. The greatest radiation of birds, at the level of order and family, occurred during the early Tertiary. Only the Passeriformes achieved a wide expansion as late as the Miocene–Pliocene. The recent species evolved during the Pleistocene. In the late Pleistocene, the estimated number of bird species was 10,600, higher than in present times.

Various orders of the Neognathae, too, have lost the ability to fly, with concurrent increases in body size. *Gastornis* and *Diatryma* of the Paleocene and Eocene of Europe and N. America, as well as the Phororhachidae (Oligocene–Pliocene) in S. America, were terrestrial predators. Penguins, limited to the southern hemisphere, had already achieved their basic adaptations to swimming and diving during the Eocene. One of the most unusual predatory birds was *Argentavis*

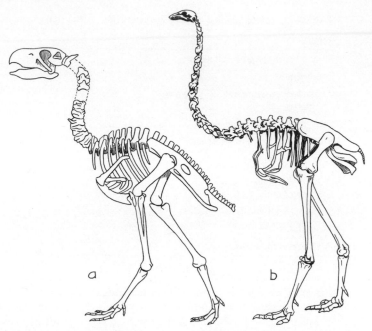

FIGURE 125. a *Diatryma* (Upper Paleocene– Lower Eocene, Diatrymiformes), restoration of skeleton, ≈2 m high. b *Dinornis* (Quaternary, New Zealand, Dinonithiformes), restoration of skeleton, over 2 m high (after *Andrews, Matthew* and *Granger*).

magnificus of the Upper Miocene of N. America, with a wingspread of 7.5 m. It is not certain whether the many avian orders are monophyletic or not.

8. Class Mammalia (mammals)

Living mammals possess diagnostic characteristics that allow a sure distinction from living reptiles (figures 126, 127).

1. Integument with hair and many glands (sebaceous, sweat, odor, lacteal).
2. Progressive evolution of the central nervous system. Cerebrum and cerebellum well developed. Twelve pairs of cranial nerves. Sense of smell originally dominant (macrosmatic).

3. Synapsid skull with two occipital condyli. Secondary palate, secondary jaw articulation, three auditory ossicles.
4. Usually seven cervical vertebrae; no free cervical and lumbar ribs; four originally pentadactylous limbs, phalangeal formula 2 3 3 3 3.
5. Dentition heterodont. Original formula of the permanent eutherian (placentalian) dentition 3.1.4.3 / 3.1.4.3, diphyodont (deciduous—milk—and permanent teeth).
6. Heart with two atria and two ventricles; three arterial stems, only the left aorta preserved.
7. Lung breathing. Thoracic and abdominal cavities divided by the diaphragm.
8. Urinary bladder, liquid urine (easily soluble urea).
9. Constant body temperature (homoiothermy).
10. Internal fertilization. Egg develops inside the mother (exception: Monotremata). Embryonic membranes (amnion, chorion, allantois). Young are nourished after birth with mother's milk. Brood care.

Distribution: Triassic–Recent.

Hair is not homologous with the reptilian scales; the question of its origin has still not been resolved. Most mammals have a keen sense of smell (are macrosmatic). Primates and marine mammals have a poor sense of smell (are microsmatic). Whales cannot smell at all (anosmatic). The prosencephalon of primitive mammals has a primarily olfactory function, yet progressive development of the neopallium, cerebral cortex, begins even in the primitive mammals. The cortex is a powerful associative centrum that takes over many functions provided in lower vertebrates by other areas of the brain. An increase in the cerebral mass and complexity is termed cerebralization. The enlarged brain also requires a larger skull cavity (cavum epeptericum), which is achieved by formation of a closed, ossified lateral wall.

TRANSITION FROM REPTILES TO MAMMALS

It is certain that mammals evolved from progressive Therapsida (mammallike reptiles). Stepwise acquisition of mammalian organization may even be traced as far back as to the Pelycosauria of the Permian/Carboniferous. Progressive changes from reptiles to mammals will be demonstrated in two examples:

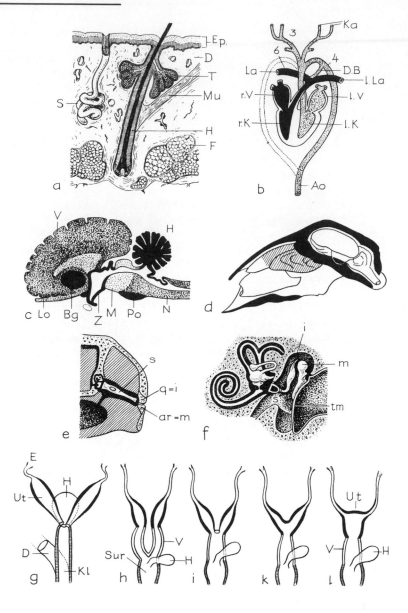

1. TRANSFORMATION OF LOWER JAW AND AUDITORY APPARATUS (figure 128)

The lower jaw of the Pelycosauria consists of six dermal bones (dentary, splenial, angular, surangular, coronoid, and prearticular) and one endochondral bone (articular). The primary joint is composed of the quadrate and the articular. During phylogeny, the size of the dentary steadily increases, while the other components become smaller. Eventually the dentary reaches to the squamosal and with it forms a secondary joint. A proof of the assumption that both joints can function simultaneously has been obtained (*Probainognathus*, Middle Triassic). As postulated by students of comparative anatomy, the primary jaw joint of the reptiles is found in the mammalian inner ear, the hammer-anvil joint. The homology of the bones is as follows:

Stapes (stirrup)	Hyomandibular of the fishes, Columella of the amphibians, reptiles, and birds
Incus (anvil)	Quadrate
Malleus (hammer)	Articular
Processus anterior of the malleus	Prearticular (gonial)
Tympanicum	Angular

This process impressively illustrates what an important role the changing functions play in the evolutionary history of animals.

2. MODIFICATION OF THE DENTITION (figure 129)

Most carnivorous reptiles devour their prey live and whole. The active mammalian way of life requires fast and complete utilization of

FIGURE 126. Body plan of mammals. **a** Section through the skin; D, dermis, Ep, epidermis, F, fat, H, hair, Mu, muscle, S, sweat gland, T, sebaceous gland. **b** Heart and aortic arches. Ao, aorta, DB, ductus Botalli, K, ventricle, Ka, carotids, La, pulmonary artery, V, atrium, 3, 4, and 6, aortic arches (phylogenic order). **c** Brain; Bg, basal ganglion, H, cerebellum, Lo, lobus olfactorius, M, mesencephalon, N, myelencephalon, Po, pons (connecting fibers of the two halves of the cerebellum), V, prosencephalon, Z, diencephalon. **d** Brain cavity and olfactory organ of a macrosmatic mammal. Olfactory mucose membrane cross-hatched. **e, f** Evolution of the middle ear and the auditory ossicles. **e** Reptile (simplified), **f** Mammal; ar, articular, i, incus (anvil), m, malleus (hammer), q, quadrate, s, stapes (stirrup), tm, tympanic membrane. **g–l** Female genital organs in mammals. **g** Monotremata. **h** Marsupialia (Didelphia). **i–l** Placentalia (Monodelphia). **i** Rodents, **k** Insectivores, carnivores, ungulates, Prosimiae. **l** Bats, apes, man; D, rectum, E, oviduct, H, urinary bladder, Kl, cloaca, Sur, sinus urogenitalis, Ut, uterus, V, vagina (after *Hildebrand, Kühn, Portmann, Romer*).

all energy that the food can provide; this is aided by mechanical processing. For this reason the mammalian dentition is usually heterodont, consisting of incisors (I, nipping teeth), canines (C, fangs), premolars (P), and molars (M, cheek teeth). The function of the molars is to grind and chew. Acquisition of the ability to masticate is

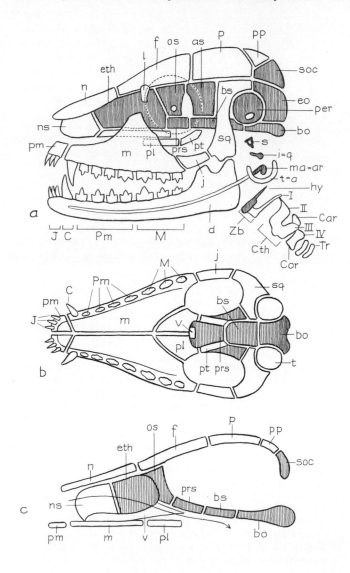

one of the decisive progressive factors in mammalian evolution. The efficient dentition of mammals exacted a price, however: reduction in the number of teeth and only one change of teeth during a lifetime. Eutheria (Placentalia) can have at most forty-four permanent teeth. The original dental formula is 3.1.4.3 / 3.1.4.3, that is, there are on each side 3 I, 1 C, 4 P, and 3 M both in the upper and lower jaws. Specialized food habits generally lead to a reduction in the number of teeth, rarely to an increase. In contrast to reptiles, mammalian teeth are changed only once during their lifetime. The formula for the deciduous (milk) tooth set is 3.1.4 / 3.1.4; only permanent dentition contains molars. The pattern of masticatory dentition had evolved during the Upper Cretaceous. A high metabolic rate in connection with the insulating hair pelt is a prerequisite of the mammalian homoiothermy (constant body temperature).

Available data on the postcranial skeletons of primitive mammals indicate that the important characteristics were already developed in certain Therapsida.

There is never any doubt whether modern forms belong to Mammalia or Reptilia; however, the distinction is not as easy where fossil finds from the Upper Triassic or Lower Jurassic are concerned. The problem is that the most significant mammal characteristics are physiological (homoiothermy, viviparity, brood care) and as such are rarely revealed in fossilized material. Therapsida of the Triassic often show a mixture of progressive (mammallike) and conservative (reptilelike) characteristics. Osteologically, the transition to mammal

FIGURE 127. Schematic drawing of a mammalian skull. **a** Lateral view. **b** Ventral view. Endochondral bones and remains of the primordial cranium are cross-hatched, dermal bones are light. Replacement (endochondral) bones of the neurocranium: eth, ethmoid, os, orbitosphenoid, as, alisphenoid, bo, basioccipital, bs, basisphenoid, eo, exooccipital, per, periotic, prs, presphenoid, soc, supraoccipital. Remains of the cartilaginous skeleton: ns, nasal septum. Dermal bones of the neurocranium: f, frontal, j, jugal, l, lacrimal, n, nasal, p, parietal, pp, postparietal (interparietal), sq, squamosal, v, vomer. Dermal bones of the viscerocranium: pl, palatine, pt, pterygoid, q, quadrate (anvil—incus). Meckel's cartilage, ends ossified: ar, articular (hammer), d, dentary. Upper jaw: m, maxilla, pm, premaxilla. From the hyoid arch are derived: hy, hyoid, s, stapes (stirrup). Branchial arch I: Zb horns of the hyoid; Branchial arch II: Cth cartilago thyreoidea (anterior part of the thyroid cartilage). Branchial arch III: Car, cartilago arytaenoidea (posterior part of the thyroid cartilage). Branchial arch IV: Cor, cartilago cricoidea (cricoid cartilage), t, tympanicum (angular), Tr, trachea. Teeth: I, incisors, C, canine, Pm, premolars, M, molars. **c** Secondary palate of the mammalian skull. Arrow marks the air passage (after *Kühn*).

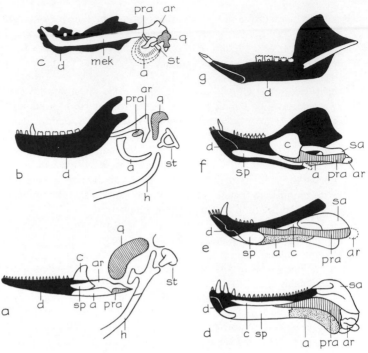

FIGURE 128. a–c Schematics to explain Reichert's theory. **a** Situation in the Reptilia. **b** In Mammalia. **c** Lower jaw and auditory ossicles of a human embryo; a, angular (tympanicum), ar, articular (hammer), c, coronoid, d, dentary (black), h, hyoid, mek, Meckel's cartilage, pra, prearticular (gonial) (dotted), q, quadrate (incus) (hatched), sp, splenial, st, stapes (columella, stir-rup). **d–f** Ordered series of synapsid lower jaws (Permian–Triassic). **d** *Dimetrodon* (Lower Permian). **e** *Cynarioides* (Upper Permian). **f** *Cynognathus* (Lower to Middle Triassic). **g** *Tritylodon* (Upper Triassic); a, angular (dotted), ar, articular, c, coronoid, d, dentary (black), pra, prearticular (hatched), sa, surangular (after *Kuhn-Schnyder*).

characteristics is gradual and continuous, so that the division between therapsids and mammals remains somewhat artificial, at least to date. Yet, it would not be sensible to deny Mammalia a class status.

MAJOR DIVISION OF THE MAMMALS (figures 126, g–h, 131)

Living mammals can be divided into three subclasses: Prototheria, Metatheria, and Eutheria. The main criterion of this division is repro-

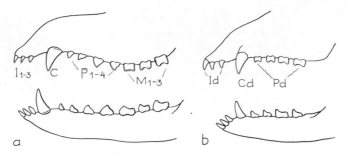

FIGURE 129. a, b Dentition of a primitive placental mammal (Eutheria). **a** Permanent dentition; I, incisors, C, canines, P, premolars, M, molars. **b** Deciduous dentition (milk teeth); Id, deciduous incisors, Cd, deciduous canines, Pd, deciduous premolars (after *Romer*).

ductive biology. The intestinal and urogenital canals of the monotremes (Prototheria) open into a common end section, the cloaca. In the marsupials (pouched mammals, Marsupialia, Metatheria) and placental mammals (Placentalia, Eutheria), the openings of urogenital and intestinal canals are separated by the perineum. Specific differences are in the structure of the female urogenital tract: in the Prototheria and Metatheria the entire female organs are duplex; in the Eutheria, the lower (distal) sections are fused. Since the three groups are not subsequent evolutionary levels, how can these subclasses be evaluated from the paleontological viewpoint? When the fossil mammalian groups are considered, Monotremata (Prototheria) lose some of their isolation. They share certain characteristics with Morganucodonta (Triassic/Jurassic), Docodonta (Upper Jurassic), and Multituberculata (Upper Jurassic–Upper Eocene). They were therefore lumped together as "Nontheria" or Prototheria sensu lato. The status of the Triconodonta (Middle Jurassic–Lower Cretaceous) is uncertain. Metatheria and Eutheria (Theria) arose in the Lower Cretaceous. Their ancestors were probably the Pantotheria, which in turn are considered to be related to Kuehneotheria or to some other, so far unknown ancestors. From the Kuehneotheria the Symmetrodonta (Upper Jurassic–Lower Cretaceous) can be further derived. Based on currently known skull characteristics, Mesozoic mammals can be grouped under two large units: Prototheria sensu lato and Theria sensu lato.

Class Mammalia

Subclass Prototheria sensu lato ("Nontheria")
  Infraclass Eotheria
    Order Morganucodonta
    Order Docodonta
    Order Triconodonta
  Infraclass Prototheria sensu stricto (Ornithodelphia)
    Order Monotremata
  Infraclass Allotheria
    Order Multituberculata (+ Haramiyidae)
Subclass Theria sensu lato
  Infraclass Trituberculata
    Order Kuehneotheria
    Order Symmetrodonta
  Infraclass Pantotheria
    Order Eupantotheria
  Infraclass Metatheria
    Order Marsupialia
  Infraclass Eutheria
    Numerous orders

1. Subclass Prototheria sensu lato ("Nontheria")

Mammals whose lateral wall of the neurocranium consists mainly of the petrosal (figure 130). Main cusps of the molars are lined up anterioposteriorly.

1. Infraclass Eotheria

Incisors not enlarged, canines present.

1. Order Morganucodonta (figure 132, a–d)

Small, unguiculate (clawed) Prototheria with primary and secondary jaw articulation. Dental formula 3–4.1.4–5.4–5 / 3–4.1.4–5.4–5. Molars with three mesiodistally ordered main cusps. Worldwide distribution at the turn of Triassic/Jurassic. *Eozostrodon (Morganucodon)*, Upper Triassic.

2. Order Docodonta (figure 132, f)

Small, omnivorous Prototheria with secondary jaw articulation. Dental formula 5. 1.3.5 / 2 + ?.1.3.5. Jurassic. Evolved probably from Morganucodonta.

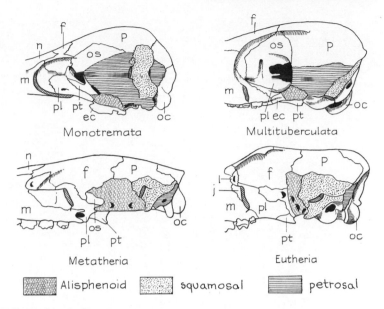

Monotremata

Multituberculata

Metatheria

Eutheria

Alisphenoid   squamosal   petrosal

FIGURE 130. Lateral wall of the mammalian braincase. In Monotremata and Multituberculata, the lateral wall of the cranium consists mainly of the petrosal, in Metatheria and Eutheria mainly of alisphenoid and squamosal. Abbreviations as in figure **127** (after *Thenius*).

3. Order Triconodonta (figure 132, e)

Small to cat-size mammals with secondary jaw articulation. Molars triconodont. Middle Jurassic–Lower Cretaceous. The exact status of this group is not yet clarified.

2. Infraclass Prototheria sensu stricto (Ornithodelphia)

Forms known to date have no incisors or canines. Molars are either present or absent.

4. Order Monotremata (monotremes) (figure 132, g, h)

Oviparous, unguiculate, plantigrade Prototheria. Jaws with horny sheaths. Teeth absent or occurring in juveniles only. Pelvic girdle with marsupial bones (ossa marsupii). There are two recent families in the Australian region: **Tachyglossidae** (echidnas) and

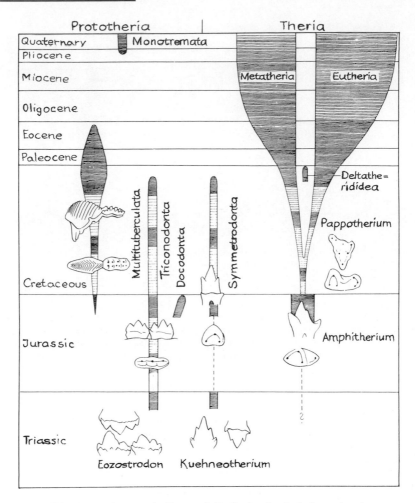

FIGURE 131. Origins of mammals. Temporal distribution (geological range) and surmised connections.

Ornithorhynchidae (platypus). Fossil history unknown. Many reptilian characteristics are retained (table 23).

Monotremata are not intermediary forms between reptiles and therians.

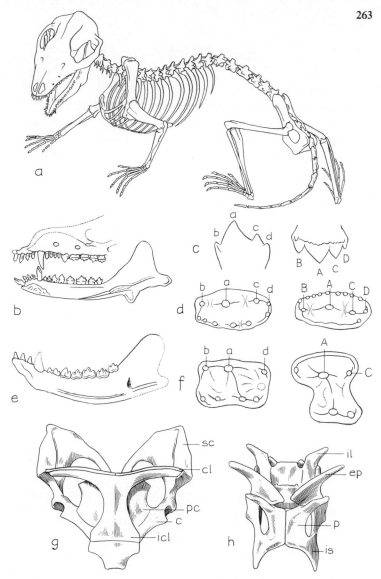

FIGURE 132. Prototheria. **a–d** *Eozostrodon* (*Morganucodon*, Upper Tertiary, Europe, Morganucodontidae). **a** Restoration of skeleton, approx. life-size. **b–d** Character of the dentition; **b** Upper and lower jaw, **c, d** Upper and lower molar, lateral view, and view of the occlusal surface. **e** *Priacodon* (Upper Jurassic, Europe, Triconodonta). Lower jaw, ≈4 cm. **f** *Docodon* (Upper Jurassic, Europe, Docodonta). Upper and lower M, occlusal surfaces. **g, h** *Ornithorhynchus* (Recent, Australia, Monotremata). **g** Pectoral girdle, ventral view; c, coracoid, cl, clavicle, icl, interclavicle, pc, procoracoid, sc, scapula. **h** Pelvic girdle, ventral view; ep, marsupial bones, il, ilium, is, ischium, p, pubis (after *Crompton, Flower, Jenkins* and *Parrington, Krusat, Simpson*).

TABLE 23. Characteristics of Monotremata

Reptilian	Mammalian
Urogenital system	Hair
Eggs with shell, abundant yolk	Mammary glands
Egg tooth in the hatching young	Differentiated smelling organ
Skull with pre- and postfrontal	Secondary jaw articulation
Stretched cochlea	3 auditory ossicles
Corpus callosum absent	Secondary palate
Free cervical ribs	Muscular cheeks and tongue
Pectoral girdle with interclavicle, procoracoid, and coracoid	Diaphragm Red blood cells without nuclei
Incomplete thermoregulation	Left aorta only

3. Infraclass Allotheria

5. Order Multituberculata (figure 133)

Mouse- to beaver-size, herbivorous Prototheria. Dental formula originally 3.1.5.2 / 1.0.4.2, reduced to 2.1.0.2 / 2.0.1.2. Upper and lower incisors are enlarged. Premolars of the lower jaw are plagiaulacoid (with sawtooth-shaped cutting edges). Molars with many cusps, ordered into two or three lengthwise rows. The plagiaulacoid dentition had evolved several times in different mammal groups independently of each other (Marsupialia, Primates). Probably only one auditory ossicle. Marsupial bones. Distribution: Upper Jurassic–Eocene. Multituberculata were a group with the greatest diversity of forms and lived the longest of all Mesozoic mammals. Three orders are distinguished by the characteristics of their dentition: **Plagiaulacoidea** (Upper Jurassic–Lower Cretaceous), **Ptilodontoidea** (Upper Cretaceous—Eocene), and **Taeniolaboidea** (Upper Cretaceous–Eocene). Multituberculata are not closely related to any of the known mammal groups.

2. Subclass Theria

Theria are mammals whose lateral wall of the neurocranium consists mainly of alisphenoid and squamosal (figure 130). The subclass con-

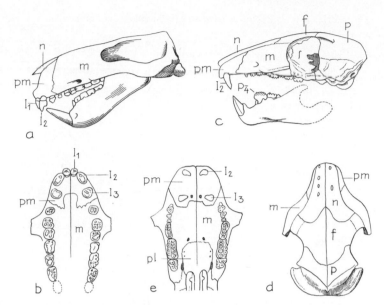

FIGURE 133. Multituberculata. **a, b** *Paul-choffatia* (Upper Jurassic–Lower Cretaceous). Skull, ≈3 cm. **b** The palate. **c–e** *Kamptobaatar* (Upper Cretaceous).

Lateral (**c**), dorsal (**d**), and ventral (**e**) views, ≈2 cm long. Abbreviations as in figure 127 (after *Clemens, Hahn, Kielan-Jaworowska*).

tains Marsupialia (infraclass Metatheria), Placentalia (infraclass Eutheria), and their fossil ancestors (infraclass Trituberculata). Living representatives are distinguished by the characteristics shown in table 24.

Considering the many common characteristics, Marsupialia and Placentalia must have had common ancestors. They differ mainly in the structure of reproductive organs, ontogeny, and type of tooth replacement. Owing to a very short gestation period, marsupialian young are born undeveloped; however, they find the maternal pouch on their own. All marsupials are born with a primary (reptilelike) jaw articulation. The true placenta of the Placentalia allows a longer gestation and higher developed newborns; this has been called the most important event of placentalian evolution. From the taxonomic point of view, the therian dentition is of special significance.

TABLE 24.   Differences between Marsupialian and Placentalian
Characteristics

Characteristic	Marsupialia	Placentalia
Braincase	Relatively small	Relatively large
Postorbital bar	Usually −	Usually +
Processus angularis, medially curved	+	−
Original dental formula	5.1.4.3 / 4.1.4.3	3.1.4.3 / 3.1.4.3
Tooth replacement	Only P4	Diphyodont
Marsupial bones	+	−
Brain hemispheres	Usually small	Increasingly large
Vagina	Duplex	Simple
Placenta	Choriovitelline	Chorioallantoid
Gestation	Short	Long
Young at birth	"Embryonal"	Precocious

THERIAN DENTITION (figures 134, 135)

One of the most elegant results of paleontological studies is proof of
fact that all of the diverse shapes of therian molars are derived from
one basic common plan. This typical molar type is designated as
tribosphenic. The tribosphenic patterns of the oldest known teeth of
both marsupials and placentals of the Upper Cretaceous are so much
alike that it is difficult to assign isolated teeth to either one or the
other group.

For descriptive purposes, the position of a tooth and its sides—in
relation to the body axes—is designated as follows:

forward = mesial, backward = distal
outward = labial (or buccal), inward = lingual

Upper and lower molars of a tribosphenic dentition are three-
edged cones, inversely positioned so that they can fit together when
the mouth is closed. Each upper and lower molar bears three cusps.

A heel (talonid) is added to the lower molar. The middle main
cusp of the upper molar is in the lingual position, the one of the
lower molar is in the labial position. In occlusion (jaws closed) the
heel of the lower molar meets the inside cusp of the next following
upper molar. There is a special nomenclature for the components of a
tribosphenic tooth. All terms that relate to the lower molars end with

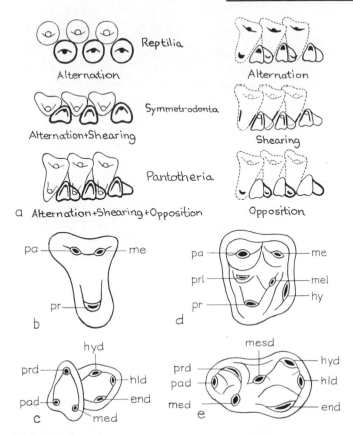

FIGURE 134. a Analysis of the occlusion of mammals. Lower teeth in bold outline. **b, c** Tribosphenic molars (M) of a primitive mammal. **b** upper M, **c** lower M. **d, e** Basic pattern of quadri-tuberculate M of early Placentalia (Eutheria). **d** Upper M, **e** lower M; end, endoconid, hld, hypoconulid, hy, hypocone, hyd, hypoconid, me, metacone, med, metaconid, mel, meta-conulus, mesd, mesoconid, pa, para-cone, pad, paraconid, pr, protocone, prd, protoconid, prl, protoconulus (after *Kuhn-Schnyder, Simpson*).

the suffix -id. The cusp triangle of the upper molar is termed trigone, the cusp of the lower molar is termed trigonid, and its heel talonid.

    Cusps of the trigone:
    pr = protocone, lingual
    pa = paracone, labial, mesial
    me = metacone, labial, distal

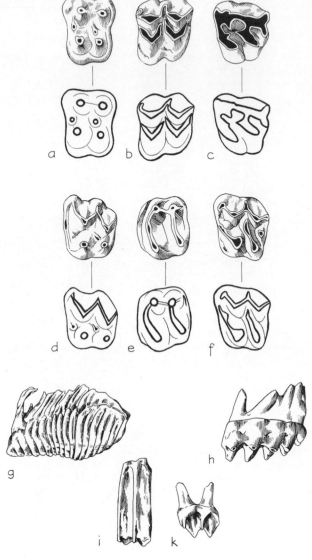

FIGURE 135. Types of mammalian upper molars: **a** Bunodont (*Hyracotherium*, Equidae), **b** Selenodont (*Protoceras*, Protoceratidae), **c** lophodont (*Rhinoceros*, Rhinocerotidae), **d** Buno-selenodont (*Palaeosyops*, Brontotheriidae), **e** Lopho-selenodont (*Tapirus*, Tapiridae), **f** Lopho-selenodont (*Anchitherium*, Equidae), **g** Hypsodont (*Elephas*, Elephantidae), **h** Brachyodont (*Gomphotherium* = *Mastodon*, Gomphotheriidae), **i** Hypsodont (*Equus*, Equidae), **k** Brachyodont (*Anchitherium*, Equidae) (after *Osborn*).

Cusps of the trigonid:
prd = protoconid, labial
pad = paraconid, lingual, mesial
med = metaconid, lingual, distal

Cusps of the talonid:
hyd = hypoconid, labial
end = endoconid, lingual

A talon with a hypocone may attach to the upper molar behind the trigone. A protoconule (prl) may occur between the paracone and protocone and a metaconule between the protocone and metacone. A hypoconulid (hld) may develop on the lower molar between the endo- and hypoconid. A lateral ridge between the crown and the root of the tooth is termed cingulum (c); vertical ridges along the tooth wall are designated as styles (style = s). No primitive structures are recognized in the deciduous dentition; it is adapted to the specialized feeding requirements of the juvenile animal.

An analysis of the function of a tribosphenic dentition shows that it can grab, shear, press, and chew. A young tooth represents only a genetically predicated "raw cast"; the full functionality of the tooth is perfected through a prolonged sharpening process. Detailed functional-morphological studies have shown that the chewing motions follow constant, fixed cycles. The tribosphenic dentition suggests that primitive Theria must have been omnivorous.

The orignal tooth grows only for a limited period and has a low crown. It is designated brachyodont. In special cases the root closes late and the crown grows longer, while being abraded on top. Teeth that compensate for wear by continuing to grow are termed hypsodont (high-crowned). The primary shape of the premolars and molars may be heavily modified. The pattern of the tooth crown changes according- ing to the feeding pattern: omnivorous, insectivorous, carnivorous, or herbivorous. While the topography of the crown of a freshly broken- through molar allows important conclusions regarding its phylogenetic status, the study of the facet pattern of a dentition reveals the feeding pattern.

The grouping of Trituberculata, Pantotheria, Metatheria, and Eutheria as the infraclass Theria is based on the tribosphenic structure of molars. The direct origin of Theria is unknown.

1. Infraclass Trituberculata

Theria with tritubercular (three-cusped) molars. Lower jaw with processus angularis.

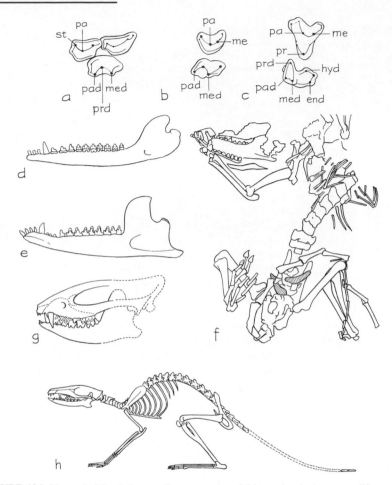

FIGURE 136. Mesozoic Mammals. **a–c** Evolution of the molars. **a** *Kuehneotherium* (Upper Triassic, Europe, Kuehneotheriidae). **b** *Amphitherium* (Middle Jurassic, Europe, Pantotheria). **c** *Pappotherium* (Upper Cretaceous, N. America, Placentalia). **d** *Peralestes* (*Spalacotherium,* Upper Jurassic, Europe, Symmetrodonta). Lower jaw, ≈3 cm. **e** *Crusafontia* (Lower Cretaceous, Europe, Eupantotheria). Lower jaw, ≈2 cm. **f** Skeleton of a Pantotherian (Upper Jurassic, Portugal), marsupial bones hatched, approx. life-size. **g** *Deltatheridium* (Upper Cretaceous, Asia, Deltatheridiidae). Skull, ≈4 cm long. Deltatheridiidae are considered to belong to a separate line, evolved independently of either Marsupialia or Placentalia. **h** *Zalambdalestes* (Upper Cretaceous, Asia, Zalambdalestidae). Restoration of skeleton, ≈30 cm, systematic status uncertain. Abbreviations as in figure **134** (after *Kielan-Jaworowska, Krebs, Krebs and Henkel, Thenius*).

1. Order Kuehneotheria (figure 136, a, b)

Very small trituberculates of which only isolated teeth and the remains of a lower jaw are known. Primary and secondary jaw articulation. One auditory ossicle (stapes). Distribution: Upper Triassic. *Kuehneotherium.*

2. Order Symmetrodonta (figure 136, d)

Shrewlike Trituberculata; only jaw and teeth remains are known. Secondary jaw articulation. Distribution: Upper Jurassic–Upper Cretaceous. *Spalacotherium,* Upper Jurassic.

2. Infraclass Pantotheria

Theria with a pretribosphenic dentition. Lower jaw with processus angularis.

Order Eupantotheria (figure 136, e, f)

Small, omnivorous or carnivorous Pantotheria with a secondary jaw joint, probably three auditory ossicles, and dentition adapted to chewing. Distribution: Middle Jurassic–Lower Cretaceous. *Amphitherium,* Middle Jurassic. *Crusafontia,* Lower Jurassic. One skeleton with marsupial bones became known from the Upper Jurassic of Portugal.

THERIA OF THE METATHERIA-EUTHERIA RANK (figure 136, c, g, h)

A heterogeneous category of Theria with tribosphenic molars, whose classification with either Placentalia or Marsupialia is uncertain. *Pappotherium,* Lower Cretaceous. This group also includes **Deltatheridiidae.** *Deltatheridium,* Upper Cretaceous.

3. Infraclass Metatheria (Didelphia)

Order Marsupialia (marsupials)

For characteristics, see figure 137.

Unguiculate, viviparous Theria with heterodont dentition. Original dental formula 5.1.3.4 / 4.1.3.4. Only P4 is replaced. General-

272

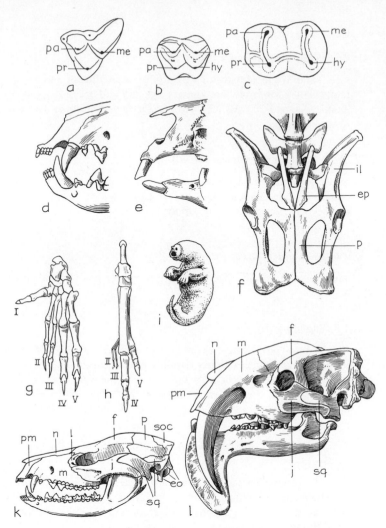

FIGURE 137. Marsupialia. **a–c** Evolution of the upper molars. **a** *Dasyurus* (Recent, Australia), carnivorous specialization. **b** *Petaurus* (Recent, Australia, glider), omnivorous, herbivorous specialization. **c** *Macropus* (Recent, Australia, kangaroo), herbivorous specialization (lophodont). **d, e** Front teeth, **d** polyprotodont (*Didelphis,* Recent, America, opossum), **e** Diprotodont (*Phascolomis,* Recent, Australia, wombat). **f** *Perameles* (Recent, Australia, bandicoot), pelvis with marsupial bones (ep). **g, h** Hind feet, **g** *Phalangista* (Recent, Australia, Phalangeridae). **h** *Macropus,* syndactyly, reduction of the Vth ray. **i** *Dasyurus,* young from the pouch, ten days old, with functional forelimbs. **k** *Didelphis,* skull, ≈8 cm. **l** *Thylacosmilus* (Pliocene, S. America, Borhyaenidae), skull, ≈20 cm. Parallel evolution with the sabertooth cats (Placentalia). Abbreviations as in figures **127, 134** (after *Elftman, Flower, Piveteau, Riggs, Romer, Young*).

ized forms have tribosphenic molars. Lower jaw with a large, in-curved processus angularis. Secondary jaw articulation, three auditory ossicles. Pubic bone with marsupial bones (ossa marsupii) in both sexes. No baculum. Macrosmatic. Corpus callosum (connection of the two halves of the brain) absent. Ureters open into a urinary bladder. Uterus and vagina are duplex, and there is a pseudovaginal canal. Mostly yolk-sac placenta. Short gestation period. Young are not fully developed at birth. They spend their juvenile days in their mother's marsupium. Distribution: Upper Cretaceous–Recent.

There was a period of explosive adaptive radiation of Marsupialia and Placentalia at the beginning of the Tertiary. As Australia became isolated from the other continents early, marsupials remained there without competition for a long time. Because habitats and niches are similar on all continents, convergent evolution of marsupials and placentals took place (marsupial mole/mole, potoroos/jerboas, gliders/flying squirrels, koala/malayan bear, thylacine/wolf). Only the habitats of large herbivores were populated differently. Marsupialia produced the kangaroos, Placentalia the ruminants. The oft-repeated view that marsupials are inferior to placentals is not altogether justified.

The dentition and the adaptations of extremities are important for the establishment of kinships among the marsupials. Many Australian marsupials show a unique specialization of the foot: the slim rays II and III are enclosed in soft parts and form a cleaning organ; this adaptation is called syndactyly. The original marsupialian dentition with numerous incisors is termed protodont. Numerous forms show a tendency toward enlarged inside incisors, in connection with the reduction of the remaining incisors, canines, and premolars. Such dentition is designated as diprotodont.

The major division of marsupials is currently subject to much discussion. The following system is based on anatomical characteristics:

Order Marsupialia

1. Suborder Didelphoidea, Lower Cretaceous–Recent, America, Europe
2. Suborder †Borhyaenoidea, Paleocene–Pliocene, S. America
3. Suborder Caenolestoidea, Paleocene–Recent, S. America, Antarctic
4. Suborder Dasyuroidea, Oligocene–Recent, Australian region
5. Suborder Perameloidea, Oligocene–Recent, Australian region
6. Suborder Phalangeroidea, Oligocene–Recent, Australian region

Present distribution of marsupials in the New World and Australia can be explained by changing paleogeographic conditions since the Upper Cretaceous. The oldest remains of Marsupialia are known from South America. So far there are none from Asia.

AMERICAN AND EUROPEAN MARSUPIALIA

1. Suborder Didelphoidea (opossums) (figure 137, d, k)

Mouse- to rat-size marsupials. Dentition polyprotodont. Dental formula 5.1.3.4 / 4.1.3.4. Molars tribosphenic. Extremities pentadactylous. Hallux divergent (grasping foot). Omnivorous, carnivorous. During the Miocene, Didelphoidea disappeared from North America and Europe. In the Pleistocene, they had spread again into North America. *Alphadon* (Upper Cretaceous, N. America), *Peratherium* (Eocene–Miocene, Europe), *Didelphis* (opossum, Pliocene–Recent, America).

2. Suborder Borhyaenoidea (figures 137, 1, 138, a)

Predominantly carnivorous Marsupialia, rat- to bear-size; there were many different forms. Dentition polyprotodont. *Eobrasilea* (Paleocene), *Thylacosmilus* (Pliocene), saber-toothed marsupial type.

3. Suborder Caenolestoidea (opossum rats) (figure 138, b)

Small, shrewlike to rat-size insectivorous marsupials. *Polydolops* (Paleocene–Eocene), P3 plagiaulacoid. *Caenolestes,* opossum rat, Recent.

AUSTRALIAN MARSUPIALIA

Representatives of the following suborders occur exclusively in the Australian region. The oldest finds are from the Upper Oligocene. By then, families were already differentiated and thereafter changed but little. A caenolestid find from the Eocene/Oligocene Antarctic suggests that Australian marsupials originated in South America.

4. Suborder Dasyuroidea (dasyurids, carnivorous marsupials) (figures 137, a, i, 138, c)

Mouse- to dog-size, carnivorous Marsupialia. Dentition polyprotodont. *Thylacinus* (thylacine, marsupial wolf, Pleistocene–Recent), *Notoryctes* (marsupial mole, Recent).

5. Suborder Perameloidea (bandicoots) (figure 137, f)

Rat- to badger-size, kangaroolike, carnivorous Marsupialia. Dentition polyprotodont. *Perameles* (bandicoot, Pleistocene–Recent)

6. Suborder Phalangeroidea (figures 137, b, c, e, g, h, 138, d, f)

Mouse to giant kangaroo size, omnivorous, predominantly herbivorous Marsupialia. Dentition diprotodont, molars quadrituberculate. Syndactyly. Tail adapted to jumping and propping the body, in some forms prehensile. Some Phalangeroidea are capable of gliding. The most form- and species-rich group of all Australian marsupials. The original group is family **Phalangeridae,** with numerous adaptations; *Petaurus* (glider, Pleistocene–Recent). Family **Phascolarctidae:** a food specialist, *Phascolarctos* (koala) feeds only on the leaves of certain eucalyptus species. Family **Thylacoleonidae** (Miocene–Pleistocene): massive skull, specialized dentition, herbivorous or carnivorous. *Thylacoleo* (Pliocene–Pleistocene) skull is as large as a lion's skull. Representatives of family **Vombatidae** (Miocene–Recent) are fossorial. *Phascolonus* (Miocene–Pleistocene), bearlike giant form. *Vombatus* (wombat, Recent). Family **Macropodidae** (kangaroos): rat- to larger than man-size marsupials with shortened forepaws and enlarged hind legs with a strong IVth toe. Dental formula 3.0–1.1–2.4 / 1.0. 1–2.4, herbivorous, partially ruminants. *Macropus* (red kangaroo, Pliocene–Recent), *Sthenurus* (Pliocene–Pleistocene), highly specialized giant form, up to 3 m high. *Diprotodon* (Pliocene–Pleistocene), rhinoceros-size contemporary of man.

4. Infraclass Eutheria (Placentalia, Monodelphia)

Originally unguiculate, viviparous Theria with heterodont dentition. Original dental formula 3.1.4.3 / 3.1.4.3. Dentition diphyodont. Molars of unspecialized forms are tribosphenic. The processus angularis of the lower jaw does not curve in. Secondary jaw articulation, three auditory ossicles. Pubic bone without marsupial bones. Baculum (os penis) often present. Macrosmatic. Mostly with corpus callosum. Ureters open into a urinary bladder. Simple vagina. Chorioallantoid placenta. Gestation period usually long. Young are born relatively precocial. Nidifugous or nidicolous. Many recent forms with cosmopolitan distribution. Temporal distribution: Cretaceous–Recent.

In North America, the yield of Eutherian fossils is excellent from Cretaceous up to Pleistocene; in Eurasia it is relatively good; in South

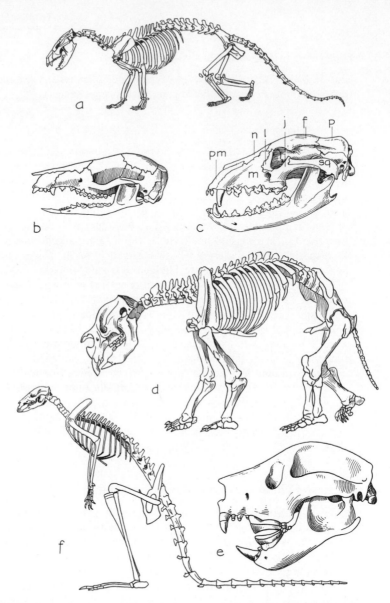

FIGURE 138. Marsupialia. **a** *Prothylacinus* (Miocene, S. America, Borhyaenoidea), skeleton, ≈1.3 m long. **b** *Caenolestes* (Recent, S. America, Caenolestoidea) skull, ≈3 cm. **c** *Thylacinus* (Recent, Australia, Dasyuroidea), skull, ≈11 cm. **d** *Diprotodon* (Pleistocene, Australia, Phalangeroidea), skeleton, up to 4 m long, giant herbivore. **e** *Thylacoleo* (Pleistocene, Australia, Phalangeroidea), skull, ≈30 cm long, shearing teeth. **f** *Macropus* (Recent, Australia, wallaby, Phalangeroidea). Abbreviations as in figure **127** (after *Gregory, Owen* and *Anderson, Parker* and *Haswell, Piveteau*).

America it is only sparse for the Cretaceous, better for the Tertiary. In Africa, finds are rare for the Paleocene but abundant since the Eocene. In Australia, Chiroptera are documented since the Miocene, murids since the Pliocene. During the Cretaceous, Eutheria hardly reached a level higher than that of a family, but since the beginning of the Tertiary up to the Lower Oligocene, there is evidence of rapid differentiation of orders and families. Then came a decline, followed by a renaissance in the early Miocene. It is assumed that Eutheria originated in Eurasia and made their way to North America and Africa. Colonization of South America is supposed to have occurred in several waves over a connection with North America. Further, it is considered that certain rodents and primates reached South America from Africa toward the beginning of the Tertiary. Eutheria reached Australia very late. Early in the Eocene, North America and Europe were connected in their northern regions, which explains their sharing of many common genera. Since the opening of the North Atlantic at the end of the Eocene about 50 million years ago, the character of mammal faunas of Europe and North America had changed.

About thirty orders of fossil and living Eutheria are recognized today. Attempts to group the orders into higher taxonomic units (cohorts) so far have been unsatisfactory. We use the following classification:

Infraclass Eutheria

1. Order Insectivora (insectivores), Cretaceous–Recent
2. Order Macroscelidea (elephant shrews), Oligocene–Recent
3. Order Dermoptera (colugos or "flying lemurs"), Paleocene–Recent
4. Order Scandentia (tree shrews) ?–Recent
5. Order Chiroptera (bats), Eocene–Recent
6. Order Primates (primates), Cretaceous–Recent
7. Order Rodentia (rodents), Paleocene–Recent
8. Order †Hyaenodonta (Creodonta, creodonts), Cretaceous–Miocene
9. Order Carnivora (carnivores), Paleocene–Recent
10. Order Lagomorpha (lagomorphs), Paleocene–Recent
11. Order †Condylarthra (condylarths), Cretaceous–Miocene
12. Order Perissodactyla (perissodactyls), Eocene–Recent
13. Order Artiodactyla (artiodactyls), Eocene–Recent
14. Order Hyracoidea (hyracoids), Eocene–Recent
15. Order Proboscidea (elephants), Eocene–Recent
16. Order Sirenia ("sea cows," dugongs, manatees), Eocene–Recent
17. Order Cetacea (whales), Eocene–Recent
18. Order †Desmostylia (desmostylians), Oligocene–Miocene
19. Order †Embrithopoda (embrithopods), Oligocene
20. Order †Notoungulata (notoungulates), Paleocene–Pleistocene

21. Order †Pyrotheria (pyrotheres), Eocene–Oligocene
22. Order †Astrapotheria (astrapotheres), Eocene–Miocene
23. Order †Tillodontia (tillodonts), Paleocene–Eocene
24. Order †Litopterna (litopterns), Oligocene–Pleistocene
25. Order †Pantodonta (pantodonts), Paleocene–Eocene
26. Order †Dinocerata (uintatheres), Paleocene–Eocene
27. Order Pholidota (pangolins), Paleocene–Recent
28. Order Tubulidentata (aardvarks), Miocene–Recent
29. Order †Taeniodonta, Paleocene–Eocene
30. Order Xenarthra (edentates), Paleocene–Recent

Xenarthra (armadillos, sloths, and anteaters) have a special status among all known Eutheria. They separated early from the other orders and flourished in South America.

EUTHERIA OF THE CRETACEOUS

Eutheria of the Cretaceous can be characterized as follows: Small, shrew- to marmot-size Eutheria with narrow, relatively long snouts. Without auditory vesicle, circular ectotympanicum. Slim lower jaw, with a medially curved processus angularis. Dental formula 5–3.1.5–4.3 / 4–3.1.5–4.3. Molars tribosphenic with high, sharp points. Two sacral vertebrae. Epipubic. Pollex and hallux not opposable. Plantigrade foot. The Cretaceous forms known to date all have a spreading manus (hand) similar to that of recent insectivores. It also allows arboreal locomotion. In the grasping hand (arboreal Primates), the carpometacarpal joint of the opposable pollex is a saddle joint.

1. Order Insectivora (insectivores) (figure 139, a–f)

Small Eutheria, mostly pentadactylous, unguiculate plantigrades. As a rule the dental formula is 3.1.4.3 / 3.1.4.3, dentition partly insectivorous, partly carnivorous. Two types of molars can be distinguished: dilambdodont molars with a W-shaped pattern, and zalambdodont molars with a V-shaped pattern. Macrosmatic. Distribution: Upper Cretaceous–Recent.

Insectivora occupy a central position among the Eutheria. The extent and division of the order are still disputed. Four suborders of the living Insectivora can be distinguished:

1. Suborder Chrysochlorida (golden moles)

Molelike. Dentition zalambdodont. Distribution: Miocene–Recent, Africa.

2. Suborder Tenrecomorpha (tenrecs)

Shrewlike and otterlike representatives. Dentition zalambdodont. Distribution: Miocene–Recent, Africa, Madagascar.

3. Suborder Erinaceomorpha

This central omnivorous group is today represented by primitive gymnures and more advanced hedgehogs. **Dimylidae** (Oligocene–Pliocene), perhaps related to the Erinaceomorpha, had adapted to a specialized diet of mollusks.

4. Suborder Soricomorpha

Great variety of forms. Soricids (shrews) and Talpids (moles) with dilambdodont molars, Solenodontids (solenodonts) with zalambdodont molars. Distribution: Eocene–Recent.

In addition, a number of fossil forms are known (**Proteutheria**), which could be regarded as the root group of the Insectivora. One well-known representative is *Diacodon* (Paleocene–Eocene). Further older Tertiary groups are **Apatemyidae** (Paleocene–Oligocene) and **Mixodectidae** (Paleocene).

2. Order Macroscelidea (elephant shrews)

Highly specialized, small Eutheria with a trunk-shaped snout, insectivorous or carnivorous (Oligocene–Recent, Africa). Fast runners and jumpers.

3. Order Dermoptera (flying lemurs, colugos)

Cat-size Eutheria with a hairy flying membrane, gliders (Paleocene–Recent). Dental formula of the recent representatives is 2.1.2.3 / 2.1.2.3; lower incisors are grooming teeth. Herbivorous and frugivorous.

4. Order Scandentia (tree shrews) (figure 139, g, h)

Squirrellike, unguiculate Eutheria. Higher degree of cerebralization than in insectivores. Dental formula 2.1.2.3 / 2.1.2.3; lower incisors form a grooming "comb." Primitive spreading hand, foot plantigrade, hallux not opposable. Omnivorous. They represent a "model" of the intermediate forms between insectivores and prosimians. Fossil documentation unattested.

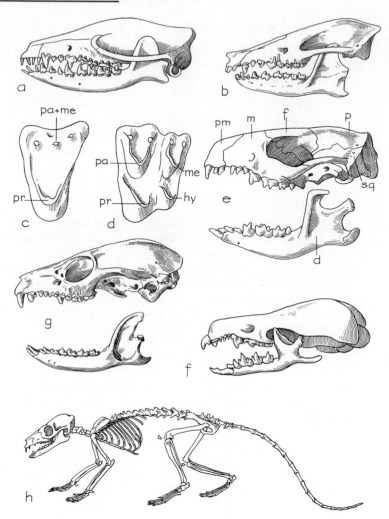

FIGURE 139. Insectivora. **a** *Kennalestes* (Upper Cretaceous), skull, ≈3 cm. **b** *Diacodon* (Lower Eocene, Lepticidae), skull, ≈6.5 cm. **c, d** Two types of upper molars. **c** *Centetes,* (Recent, Tenrecidae), zalambdodont type, **d** *Myogale* (Recent, Erinaceidae). **e** *Erinaceus* (Recent, hedgehog), skull, ≈5 cm. **f** *Myosorex* (Recent, Sori-cidae), skull, ≈2 cm. **g, h** Scandentia: **g** *Ptilocercus* (Recent, Tupaiidae), skull, ≈3.5 cm, **h** *Tupaia* (Recent, Tupaidae), skeleton, ≈20 cm long. Abbreviations as in figures **127, 134** (after *Allen, Gregory, Kielan-Jaworowska, Kuhn-Schnyder, Matthew* and *Granger, Piveteau*).

5. Order Chiroptera (bats and flying foxes) (figure 140)

From tiny to marten size, actively flying Eutheria. Dental formula
0–2.1.1–3.1–3 / 1–3.1.1–3.1–3, upper molars often dilambdodont,
lower molars tribosphenic. The flying membrane forms between the
much elongated IId and IVth fingers. Two suborders can be
distinguished:

1. Suborder Microchiroptera (bats)

Distribution: Lower Eocene–Recent. The most diverse chiropteran
group. From the original diet of insects, various genera also became
fish and crayfish catchers, predators of small vertebrates, blood suck-
ers, or fruit, nectar, and pollen eaters (pollinators). Their best
developed sense is hearing. Echolocation by ultrasound waves. Currently cosmopolitan, except the Arctic and Antarctic. *Icaronycteris*
(Lower Eocene) is a primitive representative of Microchiroptera.
Macrochiroptera evolved from this group.

2. Suborder Megachiroptera (flying foxes)

Doglike head, wingspread up to 1.5 m. Frugivores, nectar eaters, or
blood eaters. The eye is their main sensory organ. Fossil finds are
rare. Recent forms inhabit the tropics and subtropics of the Old
World. Temporal distribution: Oligocene–Recent.

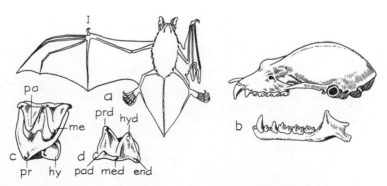

FIGURE 140. Chiroptera. **a** *Pizonyx* (Recent, Microchiroptera), showing the flying membrane. **b** *Rhinolophus* (Recent, horseshoe bat, Microchiroptera), skull, ≈3 cm. **c, d** Microchiroptera: **c** Upper molar, **d** Lower molar. Abbreviations as in figure 134 (after *Eisentraut, Friant, Miller*).

6. Order Primates (primates) (figures 141–43)

This order is of special interest because man ascended from its ranks. Primitive insectivores of the Cretaceous are considered to have been ancestors of the primates. Scandentia (tree shrews) may be regarded as a "model" of the transitional form between insectivores and primates.

The secret of the primates' success is, in addition to their adaptation to arboreal life, their retention of several primitive characteristics, particularly with respect to the extremities. Arboreal adaptations are the touch-sensitive lines on the inside of grasping-type hands and feet as well as the perfection of stereoscopic vision. There is increasing development of the neopallium. Microsmatic. The number of incisors and premolars decreases from the initial dental formula 3.1.4.3 / 3.1.4.3. Fossil finds are rare because most primates prefer tropical regions. Distribution: Upper Cretaceous–Recent.

Dentition and structure of the tympanic region are the important taxonomic characteristics. Among living primates, Lemuriformes and Lorisiformes are grouped together as Strepsirhini, having a mucous membrane strip running from the nose to the buccal cleft. Haplorhini, in contrast, do not develop this characteristic (Tarsiiformes, Platyrhini, and Catarrhini) (figure 141, b, c). Another group, the Plesiadapiformes, that had a specialized dentition, occurred from the Upper Cretaceous until the Eocene. Primates can be classified as follows:

Order Primates

1. Suborder †Plesiadapiformes, Upper Cretaceous–Eocene
2. Suborder Strepsirhini
   1. Infraorder Lemuriformes, Eocene–Recent
   2. Infraorder Lorisiformes, Miocene–Recent
3. Suborder Haplorhini
   1. Infraorder Tarsiiformes, Paleocene–Recent
   2. Infraorder Platyrhini, Oligocene–Recent
   3. Infraorder Catarrhini, Oligocene–Recent
      1. Superfamily Cercopithecoidea, Miocene–Recent
      2. Superfamily Hylobatoidea, Miocene–Recent
      3. Superfamily †Oreopithecoidea, Miocene
      4. Superfamily Hominoidea, Oligocene–Recent
         1. Family Pongidae, Oligocene–Recent
         2. Family Hominidae, Miocene–Recent

First primate radiation, **Plesiadapiformes** occurred during the older Tertiary (*Plesiadapis*). The first forerunners of Strepsirhini (*Adapis, Notharctus*) and Haplorhini (*Necrolemur, Pseudoloris*) became known during the subsequent radiation during the Eocene in North America and Europe (figure 141, a).

Strepsirhini (figure 141, d–g): Living **Lemuriformes** are limited to Madagascar and the Komor Islands. The giant *Megaladapis,* a phyllophagous, grasping climber, became extinct in historic times. Living **Lorisiformes** have discontinuous geographic distribution (Africa, South Asia).

Haplorhini: Tarsiiformes (figure 141, h, i) are represented by only one recent genus, the highly specialized *Tarsius* (tarsier, Southeastern Asia). It has old Tertiary relatives. The oldest representatives of the higher primates (**Simiae**) are known from the Oligocene of Egypt (*Parapithecus, Apidium*) and from South American (*Branisella*). Dental formula 2.1.3.3 / 2.1.3.3. Recent **Platyrhini (broad-nosed or New World monkeys)** include **Cebidae** and **Callithrichidae,** arboreal forms with long, tactile, and prehensile tails (figure 142, a, c). Dental formula 2.1.3.3 / 2.1.3.3. Their ancestors evidently drifted from Africa in the late Eocene.

Catarrhini (narrow-nosed or Old World primates) (figure 142, b, d, e, g) form a natural group and are today widely distributed over Africa and Asia. Dental formula, 2.1.2.3 / 2.1.2.3. Bilophodont lower molars are characteristic of the large and varied group of **Cercopithecoidea.** Well known is the *Mesopithecus* (Pliocene, Eurasia). In the remaining Catarrhini, a "Dryopithecus" pattern evolved from the original tribosphenic one. A Y-shaped groove is formed between five alternating main cusps. **Hylobatoidea** are today represented by the gibbons and siamangs of Southeastern Asia. They had already separated from **Hominoidea** during the Miocene. Characteristics shared with the Pongidae are parallel phenomena.

Pongidae: The largest recent primates belong to this group: gorilla (*Gorilla*), Africa; chimpanzee (*Pan*), Africa; orangutan (*Pongo*), Malayan region. The *Aegyptopithecus* (Upper Oligocene, Egypt) is regarded as an ancestral form of the pongids. Numerous forms designated as Dryopithecinae and Sivapithecinae lived in the Miocene and Pliocene of Africa and Eurasia; of these, the *Sivapithecus-Ramapithecus* group (Middle Miocene) is considered to be most closely related with the orangutan. The arboreal genus *Oreopithecus* (Upper Miocene) is regarded as a dead-end sideline of the Hominoidea.

Hominidae (figures 142, g, 143): like the pongids, the hominids also originated from the *Dryopithecus* group of forms. The parabolic dental arch is a key characteristic of a hominoid dentition; in pongids, the two rows of postcanine teeth run parallel to each other. Anthro-

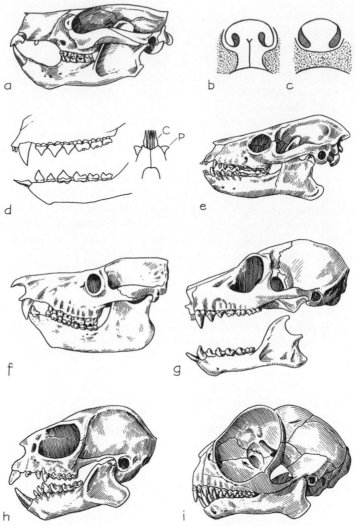

FIGURE 141. Primates. **a** *Plesiadapis* (Pleistocene, Plesiadapiformes), skull, ≈6 cm long. **b, c** Nasal area of the primates: **b** Strepsirhine type (lemurs), **c** Haplorhine type (*Tarsius*). **d** Dentition of a lemur. Lower canine (C) incisiviform, lower premolar (P) caniniform. **e** *Notharctus* (Eocene, Lemuriformes), skull, 8 cm long.

f *Megaladapis* (Quaternary, Madagascar, Lemuriformes), skull, ≈30 cm long. **g** *Galago* (Recent, Lorisiformes), skull, ≈7 cm long. **h** *Tetonius* (Eocene, Tarsiiformes), skull, ≈4.5 cm long. **i** *Tarsius* (Recent, Tarsiiformes), skull, ≈3.5 cm long (after *Gregory, Le Gros Clark, Russell, Starck*).

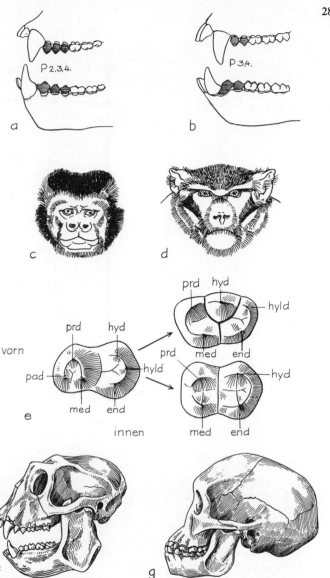

FIGURE 142. Primates. **a** Dentition of a New World monkey (*Cebus*), with three premolars (P), dental formula 2.1.3.3 / 2.1.3.3. **b** Dentition of an Old World monkey (*Macacus*) with two premolars, dental formula 2.1.2.3 / 2.1.2.3. **c** Face of a New World monkey (*Cebus*, Platyrrhini). **d** Face of an Old World monkey (*Macacus*, Catarrhini). **e** Development of the lower molar pattern of the Catarrhini. Left, the original, tribosphenic molar; upper right, *Dryopithecus* pattern (grooves forming a "Y"); lower right, bilophodonty. Abbreviations as in figure **134**. **f** *Mesopithecus* (Pliocene, Cercopithecoidea), skull, ≈8 cm long. **g** *Australopithecus* (Pliocene), restoration of the skull of a young individual (after *Ankel, Gaudry, Le Gros Clark, Romer*).

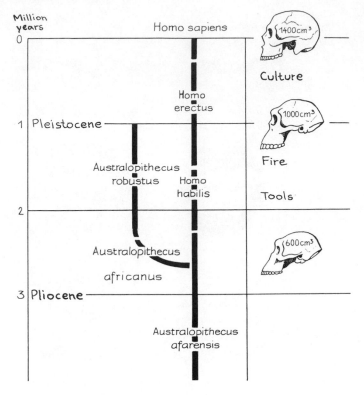

FIGURE 143. Evolution of the Hominidae

pogenesis was initiated with the acquisition of an upright walk. This evolutionary phase is designated as the transition stage between animal and man. *Australopithecus* finds from South and East Africa are characteristic of this stage. In East Africa, absolute dating is possible, thanks to the occurrence of volcanic ash, tufa, and lava. The oldest finds, *Australopithecus afarensis,* are from Tanzania and Ethiopia (3.59–3.00 million years old). The following form (in terms of time) is *Australopithecus africanus* from Omo and South Africa (two to three million years old). Its cerebral volume averages 440 cm³, and body weight is 20–30 kg. The *Australopithecus robustus* (40–80 kg), which derives from the *Australopithecus africanus* is somewhat younger; it represents a sterile sideline, ending about one million years ago. It is not positively documented whether the Australopithecinae used tools. They did not know fire. Whether *Australopithecus africanus* can be considered a direct ancestor of human beings is debated.

However, a true Hominid, *Homo habilis* (brain volume, 650 cm³, height, 1.2 m), was also discovered in East Africa. Oldest finds are circa two million years old. Found with *Homo habilis* were primitive artifacts, stone tools consciously fashioned to serve a definite purpose. With this trait *Homo habilis* passed beyond the transitional stage between animal and man. The phase that follows is represented by the *Homo erectus* group, with a widespread distribution from the Malayan Archipelago over East Asia to Europe and Africa. The skull of this form shows a characteristic combination of a small brain (700–1,200 cm³) and a large chewing apparatus. *Homo erectus* made tools and knew how to use fire. The oldest representatives (Java, E. Africa) are circa 1.5 million years old; these forms reached Europe much later. The ancestors of *Homo sapiens*, distinguished by a further increased brain (skulls from Steinheim and Swanscombe, 300–200,000 years old), occurred along with the latest *Homo erectus*. A line from such *sapiens* forms leads to the Neanderthal. "Preneanderthals" from Europe are circa 100,000 years old. The classical Neanderthaler, *Homo sapiens neanderthalensis*, lived during the first part of the Würm glacial epoch in Europe and in the Near East, 70,000–55,000 years ago. His cranial volume was 1,350–1,750 cm³. He carefully buried his dead.

Modern man, *Homo sapiens sapiens*, appeared in the Near East during the beginning of the Würm glacial epoch. During the following warmer interglacial period, he expanded across Europe and either drove out or absorbed the classical Neanderthaler. He was the "higher hunter." One of the most important events of human history was the ensuing development of the neolithic culture. This new achievement was marked by the beginning of food production (plant and animal husbandry), which predicated a settled way of life. Villages with many inhabitants were the first centers of higher culture.

The evolution of vertebrates in the Earth's history occurred in steps characterized by an increasing degree of organization and differentiation. Transition between individual levels must always bring intermediate forms with mixed characteristics. This was true of humans as well. They began with the classical mixed forms, in this case the Australopithecines. The conservative characteristics were a small brain and an apelike skull structure; a progressive trait was the upright walk. Bipedalism was a key feature of human evolution. It freed the hands for other tasks in service of higher life activities. It was followed by the reshaping of the skull, enlargement of the braincase, and reduction of the jaws. Along with this evolution came increasing independence from the natural environment. Mental faculties enabled humans fundamentally to change the ecological conditions and to exert sensory and physical influence over the whole Earth. At the same time, a great population explosion occurred. Humans ascended above the mammal class and reached a new, higher plane. By this line of reasoning, humans ought to be assigned a new class, the highest in the system of animal classification.

7. Order Rodentia (rodents) (figures 144, 145)

Generally small, unguiculate Eutheria, seldom with hoof-shaped fingernails. Dental formulas vary from 1.0.3.3 / 1.0.3.3 to 1.0.0.2 / 1.0.0.2. Only one incisor remains (I 2). It is long, curved, and has a persisting pulp. It is covered anteriorly by a layer of enamel; through wear it takes on a sharp, chisellike edge. Incisors and molars are separated by a wide diastema. Molars are brachyodont-bunodont, -lophodont, hypsodont-laminar, or prismatic with persistent pulp. Jaw articulation allows propalinal and transversal motion. Macrosmatic. The postcranial skeleton usually shows little specialization. Originally pentadactylous extremities are often plantigrade, occasionally digitigrade. Distribution: Paleocene–Recent.

Rodentia exhibit a wide range of adaptations; they are numerous and distributed worldwide (except in the Antarctic). In number, their species and genera exceed those of all other mammals. Living representatives alone comprise 32 families and 364 genera with about 1,800 species. In addition, there are many fossil forms that represent only a fraction of the once-existing species. Permanently growing incisors are their key characteristic. Repeated radiations led to parallel evolution, particularly of the cheek teeth. For this reason the classification is often difficult and unsatisfactory. With the special dentition, highly differentiated jaw musculature also evolved. Protromorphous, sciuromorphous, myomorphous, and hystricomorphous types can be distinguished by the development of the masseters. Small teeth of fossil rodents often occur in large numbers and are easily secured with proper methodology. They not only provide rich material for biometric studies but also serve as good guide fossils.

The roots of Rodentia are sought among the Insectivora, close to the primates. The oldest known find is Paleocene *Paramys;* it already shows all of the key characteristics of the order. The first radiation

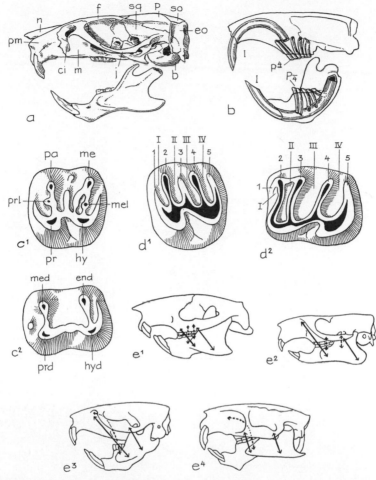

FIGURE 144. Rodentia. **a** *Rattus* (Recent, rat), skull; **b** Bulla tympanica, ci, canalis infraorbitalis. **b** *Geomys* (Recent, pocket gopher), skull, sagittal section showing the size and position of the incisors (I). **c** Nomenclature of brachyodont molars (M), **c₁** Upper M, **c₂** Lower M. **d** Nomenclature of hypsodont M, **d₁** Upper M, **d₂** Lower M; I–IV synclines, 1–5 anticlines. **e** The most important types of differentiation of musculus masseter (chewing muscle), **e₁** Protromorphic, **e₂** Sciuromorphic, **e₃** Myomorphic, **e₄** Hystricomorphic. Further abbreviations as in figures **127**, **134** (after *Bailey, Grassé, Wood*).

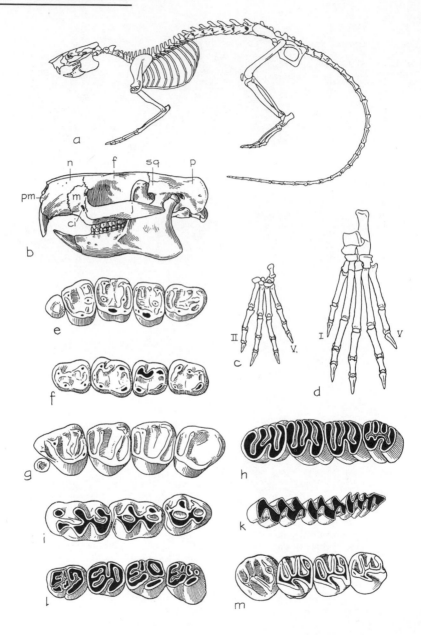

a

b

pm

n   f   sq   p

m

ci

j

e

f

c

II   V.

d

I   V

g

h

i

k

l

m

(Eocene, Protrogomorpha) was followed at the beginning of the Oligocene by a second one, of forms with an advanced type of masseter. Eight suborders are distinguished here, in keeping with E. THENIUS:

Order Rodentia

1. Suborder Protrogomorpha, Paleocene–Recent
2. Suborder Sciuromorpha, Oligocene–Recent
3. Suborder Castorimorpha, Oligocene–Recent
4. Suborder Anomaluromorpha, Oligocene–Recent
5. Suborder †Theridomorpha, Eocene–Oligocene
6. Suborder Glirimorpha, Eocene–Recent
7. Suborder Myomorpha, Eocene–Recent
8. Suborder Hystricomorpha, including Caviomorpha, Eocene–Recent

All rodents are derived from Old Tertiary **Paramyidae**. The most species-rich group is **Myomorpha**, with more than 1,000 living species. The central group of **Muroidea** are **Cricetidae**. Among the youngest descendants are **Arvicolidae** (voles) and **Muridae** (true mice) that, by "island hopping," reached even the Australian region.

Hystricomorpha are represented in both the Old and New World. They probably reached South America by the transatlantic route from Africa. The oldest known South American rodents are **Erethizontidae** (New World porcupines) and **Octodontidae** (rock rats), from which **Caviomorpha** are derived. **Hystricomorpha** (Old World porcupines) probably derived from the Eocene **Phiomyidae** (Africa). Recent Old World porcupines live in Southern Europe and in Africa. Their similarities with the New World forms result from parallel and convergent evolution.

FIGURE 145. Rodentia. **a–f** *Paramys* (Paleocene, Europe, N. America, Paramyidae, Protrogomorpha). **a** Restoration of skeleton, ≈60 cm long, **b** Skull, **ci**, canalis infraorbitalis, **c** Manus, **d** Pes, **e** Upper left cheek teeth, **f** Lower right cheek teeth. **g** *Sciurus* (Recent, squirrel, Sciuromorpha), upper left cheek teeth. **h** *Castor* (Recent, beaver, Castorimorpha), lower left cheek teeth, **i** *Cricetus* (Pliocene, Europe, Asia, Myomorpha), upper left molars. **k** *Microtus* (Recent, Europe, Asia, N. America), upper right molars. **l** *Hystrix* (Recent, Europe, Asia, Africa, Hystricomorpha), lower left cheek teeth. **m** *Erethrizon* (Recent, N. America, Hystricomorpha), upper left cheek teeth. Abbreviations as in figure **127** (after *Schaub, Stehlin* and *Schaub, Wood*).

CARNIVORES, PREDATORS

It is understandable that, with the evolution of herbivores, some of the descendants of insectivorous Cretaceous mammals turned to a carnivorous way of life. Necessary adaptations involved both body and behavior. The sense of smell was developed exceedingly well (macrosmatic animals). The most notable modifications concerned dentition. A shearing apparatus is required to eat meat. One pair of highly specialized cheek teeth (carnassials) developed into such an apparatus. The position of these breaking and shearing teeth is different in different groups. Evidently the optimal position is P 4 / M 1 (figure 146, a). The massive lower jaw possesses a tight hinge joint. Powerful jaw musculature requires large attachment areas. Both of the following orders evolved from the Insectivora of the Upper Cretaceous.

8. Order Hyaenodonta (creodonts) (figure 146, b, c, d, f, h)

Extinct, digitigrade or semiplantigrade, carnivorous, archaic Eutheria. Prominent predators of the Eocene, wolf to bear size. Carnassial pair M 1 / M 2 (**Oxyaenidae,** Upper Paleocene–Eocene) or M 2 / M 3 (**Hyaenodontidae,** Eocene–Miocene).

9. Order Carnivora ("true carnivores")

The order Carnivora can be divided into predominantly terrestrial Fissipedia and marine Pinnipedia.

1. Suborder Fissipedia (figures 146, a, e, g, 147, 148)

Large group of carnivorous or omnivorous Eutheria, whose key characteristic is the carnassial pair P 4 / M 1. Bulla tympanica ossified. Dental formula 3–2.1.4–2.4–1 / 2–1.1.4–2.4–1. Radius and ulna separate. Femur without a third trochanter. Extremities planti- to digitigrade. Pollex and hallux may be reduced. End phalanges mostly clawed, not retractable. Classification is based on the structure of the otic region. Dentition alone is insufficient to clarify phylogenetic relationships.

1. Suborder Fissipedia

1. Superfamily †Miacidea, Paleocene–Eocene
2. Superfamily Arctoidea, Eocene–Recent
3. Superfamily Cynoidea, Oligocene–Recent
4. Superfamily Aeluroidea, Eocene–Recent

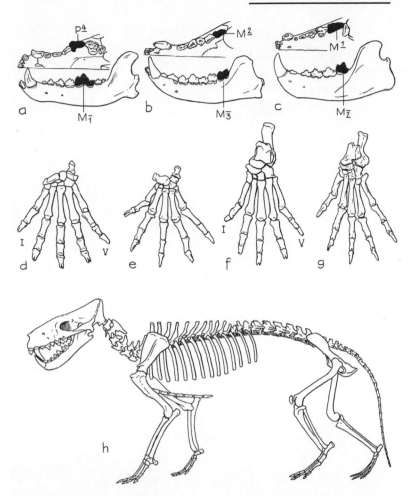

FIGURE 146. Carnivores. **a–c** Crushing/ shearing dentition of the carnivores. **a** *Canis* (Recent, Carnivora), **b** *Hyaenodon* (Eocene–Oligocene, Hyaenodonta), **c** *Oxyaena* (Upper Paleocene–Lower Eocene, Hyaenodonta), **d, f** *Oxyaena* (Upper Paleocene–Lower Eocene, Hyaenodonta), manus and pes. **e, g** *Vulpavus* (Eocene, Miacidae, Carnivora), manus and pes. **h** *Hyaenodon*. Restoration of skeleton, ≈1.2 m long (after *Denison, Matthew, Scott*).

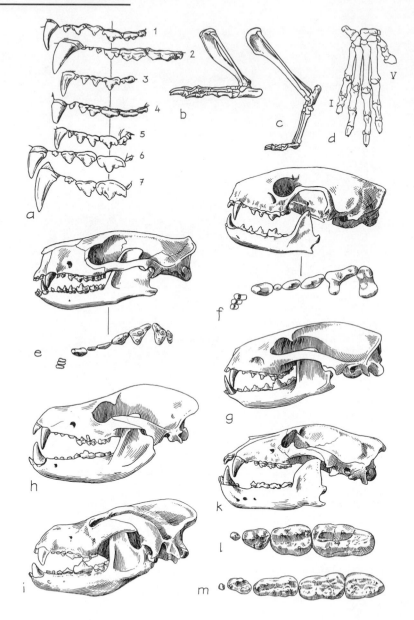

Miacidea (figure 147, e): May be regarded as ancestors of all Fissipedia.

Arctoidea (figures 147, f–m, 148, a): Families **Mustelidae** (mustelids) and **Procyonidae** (racoons) of the Arctoidea had already occurred in the Upper Eocene. The oldest known mustelid is *Amphictis* (Eocene–Oligocene, Europe). Main branches of the mustelids are documented from Oligocene–Eocene (weasel, marten, wolverine, badger, otter, skunk). Lutrinae (otters, Upper Oligocene) are regarded as the ancestral group of Phocidae (seals). Procyonidae, who have no carnassials (omnivores) and are mainly tree climbers, arrived as "island hoppers" in South America during the Upper Miocene. Today they are found only in the New World. The pandas (*Ailurus,* Asia) are an isolated group. **Ursidae** (bears), without carnassials, omnivorous, with pentadactylous, plantigrade extremities, derive from Upper Oligocene Arctoidea (*Cephalogale*). From the Holarctic bears of the Miocene evolved New World Tremarctinae and (originally) Old World Ursinae ("true" bears). During the Pliocene and the Pleistocene, Ursinae migrated to North America and supplanted the Tremarctinae, of which the last representative, the spectacled bear (*Tremarctos ornatus*), lives in South America. The cave bear (*Ursus spelaeus*) is characteristic for the younger Pleistocene. The polar bear (*Ursus maritimus*) evolved only quite recently from the brown bears. Relatives of bears are the **Ailuropodidae** (giant pandas), large herbivores (bamboo eaters), that now live in Szetschuan. From the Ursidae evolved Otariidae (eared seals, sea lions). Heavily built **Amphicyonidae** (Upper Eocene—Lower Oligocene), with reduced carnassials, are known only as fossils.

Cynoidea (figure 148, b): To Cynoidea belong the family **Canidae,** which can be derived from the *Hesperocyon* species of the Oligocene. Borophaginae (extinct in the Pleistocene) and Caninae are

FIGURE 147. Carnivora ("true carnivores"). **a** Maxilla dentition of the Fissipedia. 1, Canidae, 2, Ursidae, 3, Mustelidae, 4, Procyonidae, 5, Viverridae, 6, Hyaenidae, 7, Felidae. A line connects the P4's. **b** *Meles* (Recent, badger), plantigrade foot. **c** *Canis* (Recent, dog), digitigrade foot. **d** *Panthera* (Recent, tiger), manus. **e** *Vulpavus* (Eocene, Miacidae), skull and upper left teeth, length of the skull ≈8 cm. **f** *Martes* (Recent, marten, Mustelidae), skull and upper left teeth, length of the skull ≈7 cm. **g** *Zoodiolestes* (Miocene, Procyonidae), skull, ≈12 cm long. **h** *Amphicyon* (Miocene, Amphicyonidae), skull, ≈30 cm long. **i** *Hemicyon* (Miocene, Ursidae), skull, ≈30 cm long. **k–m** *Ursus arctos* (Recent, brown bear, Ursidae), **k** Skull, ≈30 cm long. **l** Upper P4–M3. **m** Lower P3–M3 (after *Boas, Colbert, Gaudry, Hildebrand, Riggs, Romer*).

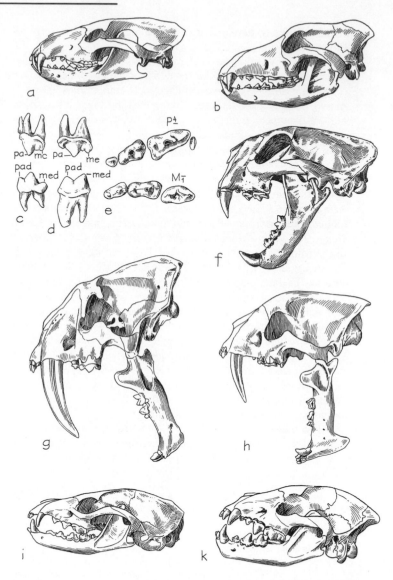

FIGURE 148. Carnivora. **a** *Daphoenus* (Oligocene, Amphicyonidae), skull, ≈24 cm long. **b** *Canis* (Recent, Cynoidea), skull. **c, d** Carnassial teeth, **c** Felidae, **d** Canidae. **e** *Felis* (Recent, Felidae), upper and lower cheek dentition. **f** *Felis* (Recent), skull, ≈24 cm long. **g** *Hoplophoneus* (Oligocene, sabretooth tiger, Felidae), skull, ≈16 cm long. **h** *Smilodon* (Pleistocene, Felidae), skull, ≈30 cm long. **i** *Genetta* (Recent, Viverridae), skull, ≈6 cm long. **k** *Hyaena* (Recent, Hyaenidae), skull, ≈20 cm long. Abbreviations as in figure **134** (after *Gregory, Jepsen, Matthew, Romer, Young*).

distinguished. Recent species (*Canis,* dogs; *Vulpes,* foxes; *Cuon,* dhols; *Nyctereutes,* raccoon dogs) arose from canid radiation during the Pliocene. The domestic dog, *Canis familiaris* derives from wolf-like forms.

Aeluroidea (figure 148, c–k): Oldest representatives (felids) are known from the Eurasian Upper Oligocene (*Eofelis, Aelurogale*).

Felidae (cats) have the most reduced and most specialized dentition of all Fissipedia. Dental formula 3.1.3–2.1 / 3.1.3–1.1(2), P 4 / M 1, carnassials. Extremities digitigrade, hand pentadactylous, foot tetradactylous. Retractable claws. Parallel characteristics make phylogenic conclusions about fossil felids difficult. Sabertooth cats developed independently of each other in three lines: *Hoplophoneus* (Oligocene), *Machairodus* (Miocene), and *Smilodon, Homotherium* (Pliocene–Pleistocene). Upper canines are long, piercing, and cutting. Small cats (*Felis*) including lynxes (*Lynx*) and cougars (*Puma*) emerged in the Upper Miocene. Panthers (*Panthera*), including jaguar, leopard, tiger, and lion, appeared as late as the early Quaternary. Cheetah (*Acinonyx*) is in an isolated position.

Representatives of **Viverridae** (civets) appeared for the first time during the Upper Eocene (*Stenoplesictis*) in Eurasia. Viverrids are primarily inhabitants of the Old World tropics and also reached Madagascar. The civets (*Viverra*) originated in Asia, genets (*Genetta*) in Africa. Mongooses (Herpestinae) have been known since the Old World Lower Miocene.

Typical **Hyaenidae** (hyenas) dentition has enlarged P 3 / P 3, which in combination with the carnassials provides a very efficient tool to aid their osteophagous feeding. Extremities digitigrade, manus usually tetradactylous, pes always tetradactylous. Claws dull, not retractable. Hyenas are descendants of viverrids, from whom they separated probably during the Oligocene. Recent spotted (*Crocuta*) and brown (*Hyaena*) hyenas belong to two separate lines.

2. Suborder Pinnipedia (pinnipeds) (figure 149)

Secondarily aquatic, microsmatic Carnivora with fin-shaped, pentadactylous extremities. Three families can be distinguished:

Suborder Pinnipedia

1. Family Otariidae, seals with ears, sea lions, Lower Miocene–Recent
2. Family Odobenidae, walrus, Miocene–Recent
3. Family Phocidae, hair seals, Miocene–Recent

Cheek teeth of **Otariidae** are simple, conical (piscivorous). In the **Odobenidae,** the upper canines are much enlarged and rootless,

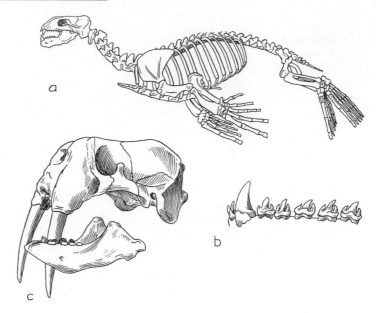

FIGURE 149. Carnivora, Pinnipedia. **a** *Arctocephalus* (Recent, fur seal, Otariidae), skeleton, ≈2.2 m long. **b** *Leptonychotes* (Recent, Phocidae), lower teeth, left. **c** *Odobenus* (Recent, walrus, Odobenidae), skull, ≈70 cm long (after *Blainville, Osborn, Weber*).

and the incisors and molars are reduced (main food is mussels). The diverse **Phocidae** have cheek teeth with one main cusp and a posterior secondary cusp (predominantly piscivorous). Otariidae evolved from primitive ursids; their place of origin is the northern Pacific. Odobenidae separated from them in the younger Tertiary. The original home of Phocidae is the Palearctic region. It is assumed that they evolved from mustelids (Lutrinae) of the Oligocene.

10. Order Lagomorpha (Duplicidentata, lagomorphs) (figure 150)

Unguiculate, herbivorous Eutheria, small to moderate in size, mostly with a dental formula 2.0.3.3 / 1.0.2.3. A small I 2 develops behind the upper I 1. The functional incisors are chisellike, continuously growing, and covered with enamel all around. Premolars and molars generally consist of two spikes with a groove in between, connected by the cementum. Cheek teeth originally brachyodont, later hypsodont and rootless. Their development from the tribosphenic pattern is subject to debate. Hand pentadactylous, foot without a hallux. Distribution: ? Paleocene, Eocene–Recent.

Characteristics in common with the Rodentia (Simplicidentata) are either primitive or result from parallel evolution. The lagomorph origin is sought among the Upper Cretaceous Insectivora. Classification relies on the morphology of the dentition.

The **Leporidae** (hares) evolved from Upper Eocene forms with brachyodont dentition. In contrast, **Ochotonidae** (pikas), which separated from Leporidae during the Oligocene, have a specialized dentition but retain a primitive postcranial skeleton. They flourished in the Miocene. Today, *Ochotona* has a Holarctic distribution. Primitive leporids appeared in East Asia during the Upper Eocene and in North America somewhat later. They separated from Archaeolaginae (*Hypolagus*). *Alilepus*, the ancestor of recent hares (*Lepus*) and rabbits (*Oryctolagus*), emerged in Europe during the Lower Pliocene. Since the Pleistocene, lagomorphs have conquered all continents. They were introduced by humans to South America and Australia.

THE "UNGULATES"

A great variety of mostly large, herbivorous, terrestrial Eutheria are included under the common terms, "ungulates" or "hoofed animals." Forms with claws are also found among them. They represent mutually independent lines with common evolutionary trends that are manifested mainly in the dentition and the organs of locomotion.

DENTITION

Primitive forms have full dentition, (3.1.4.3 / 3.1.4.3) without gaps between their teeth. Large gaps develop between the incisors, canines, and premolars through lengthening of the jaw. Incisors serve to tear out and shear off the plant; the upper incisors could be replaced by a horny pad. Canines, which had lost their function, may be reduced or absent. Lower canines may assume the form and function of a fourth incisor. Chewing efficiency of the cheek teeth depends on their surface size and resistance to wear. Thus the molars become four-cusped and reinforced by basal ridges and styles. Folding of the enamel and development of cementum makes the teeth more resistant to wear. In progressive ungulates, especially the grass eaters, cheek teeth are hypsodont and form roots only when old. This trait is most developed in the Equidae.

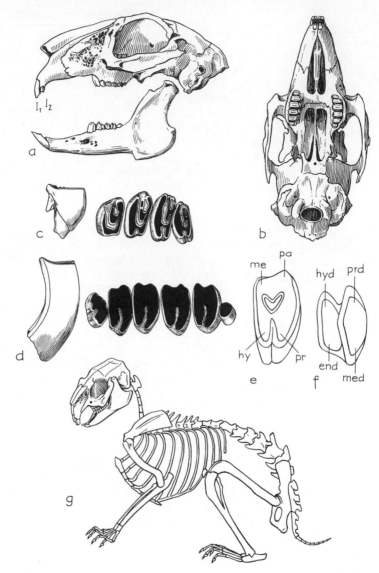

FIGURE 150. Lagomorpha. **a, b** *Lepus* (Recent, Leporidae), skull, lateral and ventral views. **c (left)** *Amphilagus* (Miocene, Ochotonidae), upper M1, subhypsodont. **c (right)** *Piezodus* (Upper Oligocene, Ochotonidae), upper P3–M2. **d (left)** *Prolagus* (Holocene, Ochotonidae), upper M1, hypsodont. **d (right)** *Hypolagus* (Pleistocene, Leporidae), upper P2–M2. **e, f** Hypothetical diagram of the lagomorph molar pattern, **e** upper M, **f** lower M. **g** *Palaeolagus* (Oligocene, Ochotonidae), restoration of skeleton, ≈20 cm long. Abbreviations as in figure **134** (after *Grassé, Tobien, Wood, Wood* and *Wilson*).

LOCOMOTION

For the herbivores, running away proved the best means of avoiding predators. Extremities of primitive ungulates are short and sturdy, pentadactylous and plantigrade, as in primitive carnivores. The digitigrade foot evolved by a shift of the metapodia (metacarpals and metatarsals) to a steeper angle, more toward the perpendicular. The unguligrade foot results when the metapodia lose contact with the ground completely. In fast locomotion, radius/ulna and tibia/fibula become longer relative to the humerus and femur, respectively, as well as to the metapodia. In addition, the number of functional toes is reduced. The IIId ray (mesaxonic type), or the IIId and IVth rays (paraxonic type) are strengthened. The ulna and fibula frequently become reduced as well. Carpalia and tarsalia, originally separate, may fuse. Heavy ungulates develop columnar limbs, with humerus and femur longer than radius/ulna or fibula/tibia, respectively. Both metapodia and spread fingers are short.

11. Order Condylarthra (condylarths) (figure 151)

Order Condylarthra unites different groups of primitive Eutheria, from weasel to bear size. They occurred primarily in the Paleocene faunas of North America and Europe. Low, elongated skull with posteriorly opened orbita. Full dentition. Brachyodont cheek teeth, mostly bunodont, often lophodont, rarely selenodont. Increasing adaptation to rapid locomotion and from omnivorous to herbivorous diet. Extremities planti- or semiplantigrade, with claws or hooves. They resemble carnivores in appearance but lack the specialized carnivore characteristics. Distribution: Cretaceous–Miocene.

Seven families have been distinguished; only three are discussed here. **Mesonychidae** (Paleocene–Lower Oligocene) probably represent the ancestral group of the Artiodactyla and the Cetacea (whales). Of the **Phenacodontidae** (Paleocene–Eocene), *Phenacodus* (N. America, Europe) is best known. It was carnivorelike in appearance but with hooves. Small, clawed *Tetraclaenodon* (Paleocene, N. America) is considered to be the ancestor of the younger Phenacodontidae and the Perissodactyla. **Didolodontidae** (Upper Paleocene–Upper Miocene) occurred in South America; they are close to the Litopterna.

12. Order Perissodactyla (Mesaxonia) (perissodactyls) (figure 152)

To the perissodactyls belong tapirs, rhinos, and equids, as well as fossil horned Brontotheria and the grotesque Chalicotheria. As gen-

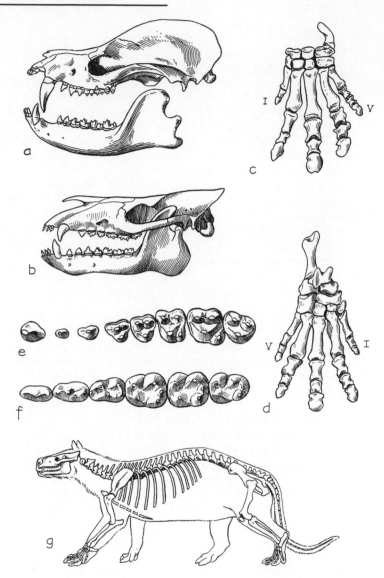

FIGURE 151. Condylarthra ("ancestral ungulates"). **a** *Deltatherium* (Middle Paleocene, Arctocyonidae), skull, ≈14 cm long. **b–g** *Phenacodus* (Lower Eocene, Phenacodontidae), **b** Skull, ≈16 cm long, **c** Left manus, **d** Right pes, **e, f** Dentition, **e** Upper right, **f** Lower left, **g** Skeleton with an outline of the body, ≈1.5 m long (after *Cope, Gregory, Matthew, Scott*).

eralized representatives of Perissodactyla, the Lower Eocene species of the genus *Hyracotherium (Eohippus)* come to the fore. The size of a fox, a *Hyracotherium* had slim limbs with four rear and three front functional toes, of which the IIId toe is strongest (mesaxony). Full dentition with brachyodont, bunodont molars. Later the molars become lophodont, premolars become molariform. Cheek teeth may be hypsodont and reinforced with dental cementum. The absence of the clavicle and the occurrence of a third trochanter on the femur are further common marks of Perissodactyla. Horn-shaped structures are found only on the skulls of certain Brontotheria and rhinoceroses. Perissodactyls flourished during the older Tertiary; in the latter part of the Tertiary they were gradually replaced by the Artiodactyla. Specific combinations of characteristics allow us to distinguish the following three suborders:

Order perissodactyla

1. Suborder Ceratomorpha (tapirs and rhinoceroses), Eocene–Recent
2. Suborder Hippomorpha (equids and Brontotheria), Eocene–Recent
3. Suborder †Ancylopoda (Chalicotheria), Eocene–Pleistocene

1. Suborder Ceratomorpha

Ceratomorpha encompass two superfamilies, Tapiroidea and Rhinocerotoidea. Both can be traced to a common ancestral group (*Homogalax,* Lower Eocene, N. America, Asia).

1. Superfamily Tapiroidea (tapirs) (figures 152, e, i, 153, a, b)

The most primitive Perissodactyla, in size from hare to rhinoceros. Upper molars bunoselenodont, lower molars lophodont. Rear feet have four-digit toes, front feet have three-digit toes, equipped with plantal pads and hooves. The brain does not achieve the developmental level of the horse brain. Great variety of Eocene forms, fast runners as well as forms of colossal size, without the specialized adaptations of modern representatives (e.g., trunk or dentition). The line of modern tapirs began with *Protapirus* (Oligocene, Eurasia, N. America). During the Pleistocene tapirs arrived in South America. Recent species are known from southern Asia (*Tapirus indicus*), Central America (*Tapirus bairdi*), and South America (*Tapirus terrestris, Tapirus pinchaque*); they provide typical examples of disjunct distribution.

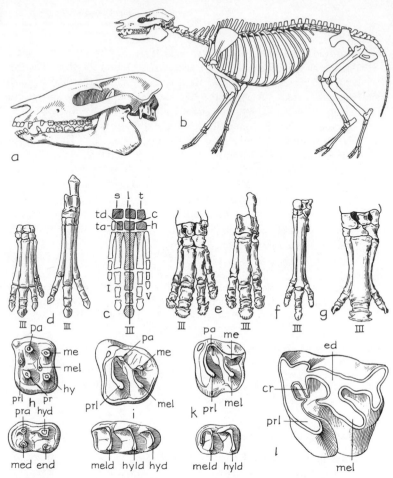

FIGURE 152. Perissodactyla. **a, b** *Hyraco-therium* (*Eohippus,* Lower Eocene, Europe, N. America, Equidae), **a** Skull, **b** Skeleton, ≈45 cm long. **c** Diagram of the mesaxonic structure of the right manus of a perissodactyl; c, capitatum, h, hamatum, l, lunatum, s, scaphoid, t, triquetrum, ta, trapezium, td, trapezoid. **d** *Hyracotherium,* manus and pes. **e** *Tapirus* (tapir, Tapiridae), manus and pes. **f** *Triplopus* (Middle Eocene, N. America, Hyracodontidae), manus of a fleet-footed rhinoceros. **g** *Baluchitherium* (Oli-gocene–Miocene, Asia, Rhinocerotidae), manus of a large, heavy rhinoceros. **h** *Hyracotherium,* upper and lower molars. **i** *Protapirus* (Oligocene, N. America, Tapiridae), upper and lower molars. **k** *Hyrachyus* (Eocene, N. America, Helaletidae), upper and lower molars. **l** *Rhinoceros* (Recent, Rhinocerotidae), upper molar; cr, crista, ed, ectolophe, mel, metalophe, prl, protolophe, further abbreviations as in figure **134** (after *Cope, Gregory, Osborn, Romer, Scott, Weber*).

2. Superfamily Rhinocerotoidea (figures 152, f, g, k, l, 153, c–i)

Perissodactyla with a tendency to phylogenic increase in size. Skull may bear one or two keratinous horns (without bony apophyses). Front dentition variously differentiated, molars lophodont. Flourished during the Eocene–Oligocene. Four families are distinguished:

1. Family †Hyracodontidae, Upper Eocene–Oligocene, N. America, Asia
2. Family †Amylodontidae, Middle Eocene–Lower Oligocene, Holarctic
3. Family Rhinocerotidae ("true" rhinos), Oligocene–Recent
4. Family †Indricotheriidae, Oligocene–Miocene, Eurasia

Rhinocerotidae are a diverse family, divided into several branches that were separated as early as the Upper Miocene. The Sumatran rhinoceros (*Dicerorhinus sumatrensis*), which carries two horns, alone remains of the *Dicerorhinus* line. The genus *Coelodonta*, to which belonged *Coelodonta antiquitatis*, a plains form from cold regions, lived during the Pliocene–Pleistocene. The only recent living representatives of unicorn rhinoceroses are the Indian rhinoceros (*Rhinoceros unicornis*) and the Java rhinoceros (*Rhinoceros sondaicus*). Black rhinoceros (*Diceros bicornis*) and white rhinoceros (*Ceratotherium simum*), in Central and South Africa, survived from the African line. Elasmotheriinae included very large, specialized, plains rhinoceroses, of which the last form, *Elasmotherium* in Eurasia, survived even the Pliocene–Pleistocene transition.

Indricotheriidae (Baluchitheriidae) (Oligocene–Miocene, Eurasia) were long-legged, long-necked giant rhinoceroses that had no horn. They stood up to 6 m tall at the withers.

2. Suborder Hippomorpha

Although both superfamilies, Equoidea and Brontotheroidea, derive from *Hyracotherium* (Lower Eocene), they went their separate ways early. Brontotheroidea rapidly increased in size and assumed the ecological role of rhinoceroses but became extinct during the Middle Miocene. Equoidea specialized and evolved into fast, graminivorous runners.

1. Superfamily Equoidea

Equoidea comprise Palaeotheriidae and Equidae.

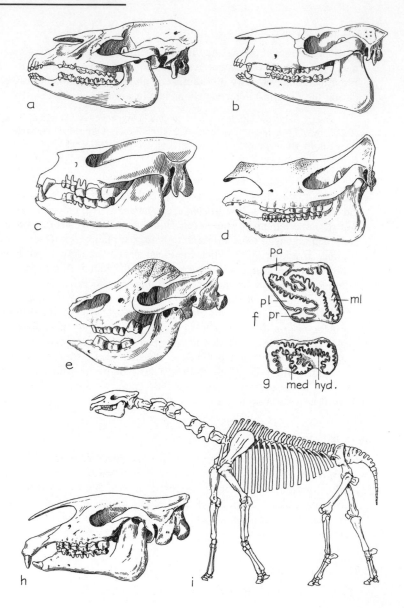

pa

pl

pr

f

ml

g   med hyd.

a

b

c

d

e

h

i

1. Family Palaeotheriidae (figure 154, a–c)

Relatively large Equoidea, reaching the size of small rhinoceroses (Middle Eocene–Lower Oligocene). Skull similar to that of tapirs.

2. Family Equidae (figures 152, a, b, d, h, 154, d–q, 155)

A group of Perissodactyla including numerous species that, during their evolution, acquired the following characteristics:

1. Increase in size.
2. Lengthening of the distal parts of extremities. Reduction of ulna and fibula. Reduction of the side toes of both manus and pes, in combination with the strengthening of the IIId toe.
3. Lengthening of the face parts of the skull.
4. "Molarization" of premolars, hypsodont cheek teeth. Complicated pattern of the tooth crown and build-up of cementum.
5. Development of the postorbital bar.
6. Increased brain capacity (size and complexity).

In different lines these characteristics developed neither simultaneously nor at the same rate. Equid phylogeny is a good example of the mosaiclike course of evolution, although during certain periods it was straightforward (linear, orthogenetic).

There are three equid subfamilies:

1. Subfamily Hyracotheriinae

Small brush forms. (*Hyracotherium*, Lower Eocene, Europe, N. America). *Orohippus* (Middle Eocene), *Epihippus* (Upper Eocene). Since the Middle Eocene, equine phylogenic development took place exclusively in North America. From there, radiations

FIGURE 153. Ceratomorpha. **a** *Protapirus* (Oligocene, N. America, Tapiridae), skull, ≈30 cm long. **b** *Hyrachyus* (Eocene, N. America, Helaletidae), skull, ≈30 cm long. **c** *Metamynodon* (Oligocene, N. America, Amylodontidae), skull, ≈80 cm long. **d** *Dicerorhinus* (Oligocene, Europe, Rhinocerotidae), skull, ≈60 cm long. **e–g** *Elasmotherium* (Pleistocene, Europe, Asia, Rhinocerotidae), **e** Skull, ≈75 cm long, **f** Upper M; ml, metalophe, pa, paracone, pl, protoloph, pr, protocone. **g** Lower M; hyd, hypolophid, med, metalophid. **h** *Baluchitherium* (Oligocene–Miocene, Asia, Rhinocerotidae), skull, ≈1.35 m long. **i** *Indricotherium* (Oligocene, Asia, Rhinocerotidae), a giant rhinoceros, skeleton, ≈5.5 m long (after *Brandt, Gromova, Hatcher, Osborn, Scott*).

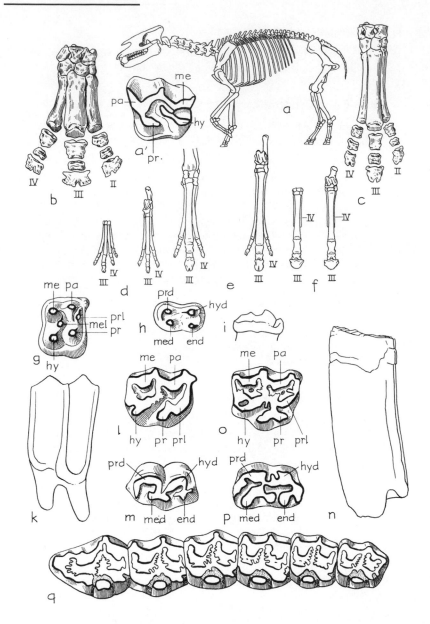

periodically occurred: to Eurasia, Africa, and the last one to South America.

2. Subfamily Anchitheriinae

From *Mesohippus* and *Miohippus* of the Oligocene gradually evolved the *Anchitherium* group. Manus functionally tridactylous, with weak side toes. Distal phalanges with hooves. Several new Miocene genera belong to two main lines. The *Anchitherium* line remains conservative; its representatives crossed the Bering Strait land bridge to Asia and migrated in Middle Miocene to Europe. The *Parahippus* line led to the Equinae in North America.

3. Subfamily Equinae

The arid conditions of the North American Miocene decimated deciduous forests and led to an open landscape, with steppe grasslands. Representatives of the *Merychippus* group evolved hypsodont dentition with dental cement that enabled them to disintegrate the silica-containing grasses. Originally brachyodont foliage eaters changed into hypsodont grass eaters. Two groups of forms separated from the *Merychippus* group: tridactylous hipparions and monodactylous equids. *Hipparion* finds from the Lower Pliocene of Eurasia and Africa are numerous. *Hipparion* became extinct during the Pleistocene. The geologically oldest single-hoofed form is *Pliohippus* of the North American Pliocene. With the genus *Equus,* a new invasion from North America spread over Eurasia all the way to Africa. Equids have survived to the present day in Asia (wild horses, kiang, onager) and Africa (zebras, wild ass). Descendants of *Pliohippus* entered the Pleistocene in the forms of *Hippidion, Onohippidium,* and

FIGURE 154. Hippomorpha. **a** *Palaeotherium* (Upper Eocene–Lower Oligocene, Europe, Palaeotheriidae), skeleton, ≈1.6 m long, **a'** Upper left M. **b** *Palaeotherium,* manus. **c** *Plagiolophus* (Upper Eocene–Lower Oligocene, Europe, Palaeotheriidae), manus. **d** *Hyracotherium (Eohippus)* (Lower Eocene, Europe, N. America, Equidae), manus and pes. **e** *Mesohippus* (Lower–Upper Oligocene, N. America, Equidae), manus and pes. **f** *Equus* (Recent, Equidae), manus and pes. **g–i** *Hyracotherium,* **g** Upper M, **h, i** Lower M. **k–m** *Merychippus* (Middle Miocene–Lower Pliocene, Equidae), upper and lower M. **n–p** *Equus* (Recent, Equidae), upper and lower M. **q** *Hipparion* (Upper Miocene, Pleistocene, Europe, Africa, Asia, N. America, Equidae). Upper P2–M3. Abbreviations as in figure 134 (after *Abel, Gaudry, Lull, Osborn, Stehlin*).

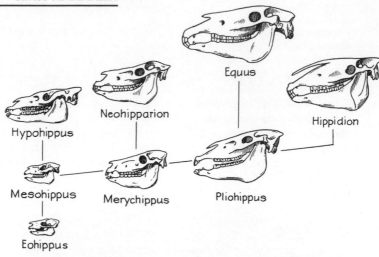

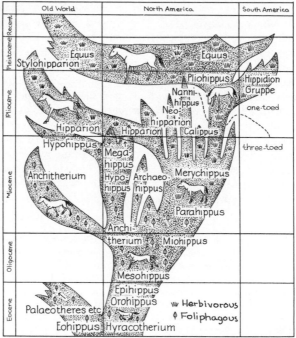

FIGURE 155. Equid evolution. *Above:* Evolution of several skulls. *Below:* Phylogenetic tree, habitat, and geo- graphic distribution. (Simplified; the course of phylogeny is in fact more complex) (after *Simpson*).

Parahipparion (S. America). However, New World equids became extinct at the latest in the early Holocene. Mustangs are feral domestic horses. The domestication of the horse, "la plus noble conquête de l'homme," probably took place in Asia. In conclusion, it should be emphasized that the characteristic Old World sequence, *Anchitherium* (Miocene), *Hipparion* (Pliocene), and *Equus* (Quaternary), does not represent phylogenetic descendancy but rather a series of different stages.

2. Superfamily Brontotheroidea (Titanotheroidea) (figure 156)

Hippomorpha with variable skull structure and a conservative postcranial skeleton. A number of lines that independently developed paired, skin-covered horns ("pseudohorns") diverged from the small, lightfooted, Lower Eocene species with no horns during the Middle Eocene. They flourished mainly during the Upper Eocene. Found among the latter were giant forms (*Brontotherium,* Lower Oligocene, N. America, over 4 m long).

3. Suborder Ancylopoda

Superfamily Chalicotheroidea (figure 157)

Aberrant, fossil Perissodactyla, from very small to very large. Cheek teeth brachyodont, upper molars bunoselenodont, lower molars lophodont. Manus and pes of later forms are highly specialized, distal phalanges deeply split. Oldest representative (*Palaeomoropus,* Lower Eocene, N. America) resembles *Hyracotherium.* By the Oligocene, two lines had become established in Eurasia. **Schizotheriini** were quadrupedal runners that also reached North America during the Miocene (*Moropus*). Large **Chalicotheriidae,** with long front limbs, appeared during the Eurasian Miocene. Dental formula 0.0.3.3 / 3.1.3.3. Manus and pes tridactylous, with powerful claws. They grasped branches of trees with their hooked front limbs to reach the tender leaves. *Chalicotherium* disappeared from Europe at the end of the Pliocene.

13. Order Artiodactyla (Paraxonia, artiodactyls) (figure 158)

The macroclassification of Artiodactyla is not unified. Although the living representatives can readily be divided into three natural groups, namely, nonruminants, camels, and ruminants, the numerous fossil

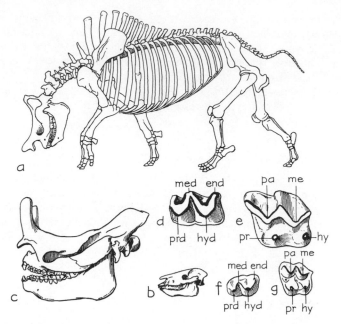

FIGURE 156. Brontotheroidea. **a** *Brontops* (Oligocene), skeleton, ≈4 m long. **b** *Eotitanops* (Lower Eocene, N. America), skull, ≈30 cm long. **c–e** *Brontotherium* (Oligocene, N. America), **c** Skull, ≈80 cm long, **d** Lower M, **e** Upper M. **f, g** *Eotitanops,* **f** Lower M, **g** Upper M. Abbreviations as in figure **134** (after *Osborn, Scott*).

forms present difficulties. The following adaptations appeared during artiodactylan phylogeny:

1. Size increase.
2. In good runners: lengthening of the distal parts of extremities, reduction of ulna and fibula, reduction of the side toes of both manus and pes, strengthening of the IIId and IVth toes, as well as fusion of the IIId and IVth metapodia into a cannon bone (paraxony). Modifications of carpus and tarsus.
3. Modification of front teeth. Transition from bunodont to selenodont molars. Development of hypsodont teeth and production of cementum.
4. Formation of a postorbital bar.
5. Production of antlers or horns.

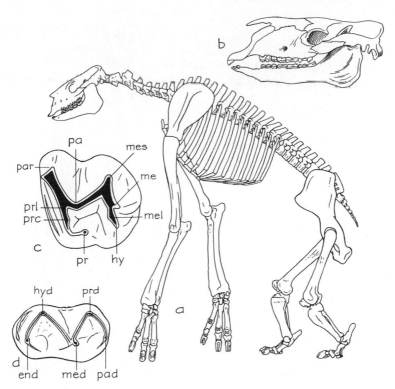

FIGURE 157. Ancylopoda. **a** *Chalicotherium grande* (Miocene, Europe, Chalicotheriidae), skeleton (usual posture), up to 2 m high. **b** *Moropus* (Miocene, N. America, Asia, Chalicotheriidae), skull, ≈60 cm long. **c, d** *Chalicotherium*, **c** Upper M, **d** Lower M; mel, metalophe, mes, mesostyle, par, parastyle, prl, protolophe, further abbreviations as in figure **134** (after *Gregory, Zapfe*).

6. Evolution of rumination in certain herbivores.

7. Increased brain capacity (size and complexity).

Distribution: Eocene–Recent. Artiodactyla currently enjoy their period of greatest prosperity.

The structure of the astragalus, a key characteristic of Artiodactyla that has already been mentioned, allows for great sagittal mobility of the hind limb. Clavicle usually absent. Femur without a third trochanter.

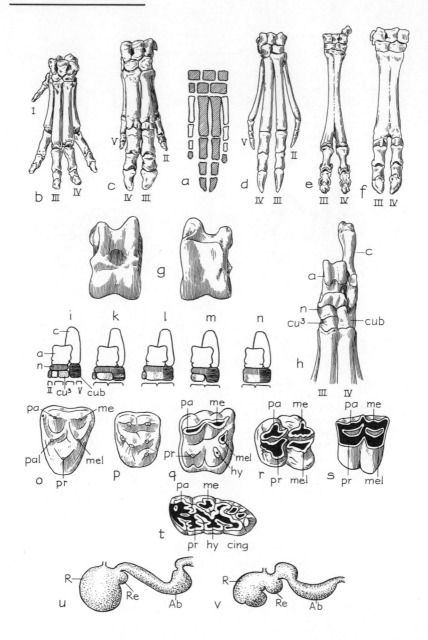

We employ the following classification:

Order Artiodactyla

1. Suborder †Palaeodonta, Lower Eocene–Middle Miocene
2. Suborder Suina, Lower Eocene–Recent
3. Suborder †"Ancodonta," Middle Eocene–Pleistocene
4. Suborder †Oreodonta, Upper Eocene–Pliocene
5. Suborder Tylopoda, Upper Eocene–Recent
6. Suborder Pecora, Upper Eocene–Recent

The roots of Artiodactyla are probably among the Mesonychidae (Condylarthra).

1. Suborder Palaeodonta (figure 159, a)

A group of six families that exhibit a mixture of archaic and specialized characteristics. They are not, however, a natural group.

2. Suborder Suina

Suina are divided into Entelodontoidea, Suoidea (pigs), and Hippopotamoidea (hippopotamuses). **Entelodontidae** (Eocene–Miocene, N. America, Europe) had a powerful, elongated facial skeleton (figure 159, b, c).

1. Superfamily Suoidea (figures 158, c, h, t, 159, d, e)

The most primitive living artiodactyls, with a full or slightly reduced dentition, molars (neo)bunodont, canines mostly strong. Omnivores.

FIGURE 158. Artiodactyla. **a** Diagram of the paraxonic structure of the right manus of artiodactyls. **b** *Brachyodus* (Oligocene–Miocene, Anthracotheriidae), manus. **c** *Sus* (Recent, Suidae), manus. **d** *Hyemoschus* (Recent, Tragulidae), manus. **e** *Camelus* (Recent, Camelidae), manus. **f** *Bos* (Recent, Bovidae), manus. **g** *Rangifer* (Recent, Cervidae), astragalus, left anterior view, right posterior view. **h** *Sus,* tarsus dorsal; a, astragalus, c, calcaneus, cub, cuboid, cu³, cuneiform 3, n, navicular. **i–n** Artiodactyl tarsus, **i** Primitive tarsus, **k** typical Tylopoda and Pecora, **l** Tragulina, **m** *Giraffa,* **n** *Okapia*. **o–t** Upper M, **o** Diagram, **p** *Dichobune* (Eocene–Oligocene, Palaeodonta), **q** *Anthracotherium* (Eocene–Miocene, Hippopotamoidea), **r** *Hippopotamus* (Recent, Hippopotamidae), **s** *Tragocerus* (Miocene–Pleistocene, Bovidae), **t** *Dicoryphochoerus* (Pliocene, Suidae); cing, cingulum. **u, v** Ruminant stomach, **u** Tylopoda, **v** Pecora; Ab, abomasum, R, rumen, Re, reticulum (after *Carlsson, Falconer, Gaudry, Grassé, Gregory, Romer, Sinclair, Starck, Viret, Wortman*).

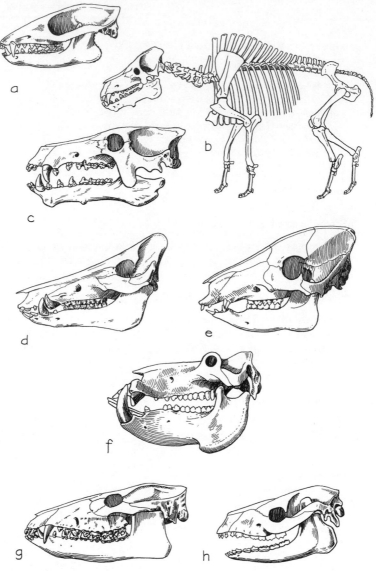

FIGURE 159. Palaeodonta, Suina, Ancodonta. **a** *Homacodon* (Eocene, N. America, Palaeodonta, skull, ≈8 cm long. **b** *Dinohyus* (Miocene, N. America, Entelodontidae), skeleton, ≈3 m long. **c** *Archaeotherium* (Oligocene, N. America, Entelodontidae), skull, ≈40 cm. **d** *Sus scrofa* (Recent, wild boar, Suidae), skull, ≈40 cm. **e** *Dicotyles* (Recent, peccary, America, Dicotyli-dae), skull, ≈20 cm. **f** *Hippopotamus* (Pleistocene, Europe, Asia, Africa, Hippopotamidae), skull, ≈60 cm. **g** *Microbunodon* (Oligocene, Europe, Anthracotheriidae), skull, ≈20 cm. **h** *Anoplotherium* (Eocene, Europe, Anoplotheriidae), skull, ≈35 cm (after *Blainville, Gregory, Hünermann, Peterson, Reynolds, Scott, Sinclair, Weber*).

During the younger Tertiary they had already separated into two distinct lines: Suidae, with modification of skull and dentiton, and Dicotylidae, with modified limbs.

Family Suidae

Their ancestral group are the Hyotheriinae (Oligocene, Europe), from which a number of younger Tertiary lines are derived. The important genus *Sus* provided the original form of the domestic pigs (wild boar, *Sus scrofa*).

Family Dicotylidae (Tayassuidae, peccaries)

Progressive structure of the limbs, median metapodia may be fused (cannon bone). Peccaries have been known since the Middle Oligocene; their evolution took place mainly in North America. *Dicotyles* is a recent New World genus with two species.

2. Superfamily Hippopotamoidea

Family Hippopotamidae (hippopotamuses) (figures 158, b, q, r, 159, f, g)

The only living artiodactyls with an amphibious way of life. Modified skull and dentition. Dental formula 3–2.1.4.3 / 3–1.1.4.3. Older species had three I on each side (hexaprotodont dentition). Tetraprotodont or, properly, diprotodont dentition resulted from the reduction of the lateral incisors. Molars with four plicated cusps (they wear to a trefoil pattern). The oldest hexaprotodont representatives are known from the South European Miocene. Dwarf forms evolved from tetraprotodont hippopotamuses (Pleistocene, Crete, Cyprus, Sicily). Recent: hippopotamus (*Hippopotamus amphibius*) and pygmy hippopotamus (*Choeropsis liberiensis*), Africa. Phylogenic origin of the hippopotamuses is not yet clarified.

3. Suborder "Ancodonta"

A heterogeneous fossil group.

1. Superfamily Anoplotheroidea

Family Anoplotheriidae (figures 159, h, 160, d, e)

> Small to moderate-sized artiodactyls of the European Eocene and Oligocene.

2. Superfamily Cainotheroidea (figure 160, a–c)

> Small artiodactyls resembling hares. Eocene to Miocene of Europe.

4. Suborder Oreodonta (figure 160, f, g)

> Fossil artiodactyls with a large, varied family, **Oreodontidae** (Lower Oligocene–Pliocene), and a small specialized family, **Agriochoeridae** (Middle Eocene–Middle Oligocene). Many oreodontid forms roamed the wide plains of North America in large herds during Oligocene–Miocene.

5. Suborder Tylopoda (camels)

1. Superfamily Xiphodontoidea

Family Xiphodontidae

> Fossil artiodactyls with didactylous limbs, metapodia separate (Eocene–Oligocene, Europe).

2. Superfamily Cameloidea (figures 158, e, 160, h–k)

> Small to very large artiodactyls without horns, currently represented by the Old World camels and South American llamas. Since the Miocene, the metapodia III and IV have grown together, with distal ends diverging. Cameloidea developed rumination independently of the Pecora. Cameloidea began as small forms during the Eocene (**Oromerycidae**, N. America). **Camelidae** evolved during the Oligocene, then separated into various branches during the Miocene. The main branch led to large llamalike forms (Pliocene). At that time *Procamelus* emigrated to Eurasia, where it subsequently evolved into the genus *Camelus*. *Camelus* split during the Lower Pleistocene into the one-humped *Camelus dromedarius* (S. Asia, N. Africa) and the two-

humped *Camelus bactrianus* (Eurasia). Both species were domesticated. Ancestors of the genus *Lama* reached South America during the Pleistocene and were also domesticated. North American camels expired with the *Camelops* by the beginning of the Holocene.

3. Superfamily Protoceratoidea

Family Protoceratidae (figure 160, l)

Strictly North American, fossil artiodactyls (Oligocene–Pliocene). Some forms with bizarre, ossified horn structures developed from forms without horns.

6. Suborder Pecora

The most varied and species-rich group of artiodactyls. The dermal bones of the skull often bear antler or horn structures, primarily in males. Hornless and antlerless forms may have enlarged upper canines. Dental formula 3.1.3.3 / 3.1.4.3; however, upper incisors and canines are mostly absent. Lower canines become incisor-shaped and join the incisors. Cheek teeth brachyodont to hypsodont, molars selenodont. Extremities generally unguligrade (hoof-bearing). Cannon bones (fused metapodia III and IV). Very efficient utilization of food is achieved by rumination. Distribution: Upper Eocene–Recent.

Pecora are customarily divided into four superfamilies:

1. Superfamily Traguloidea (figure 158, d)

Small, hornless, primitive Pecora, rare in the Upper Eocene, more numerous in the Oligocene. Today they are represented by *Hyemoschus* (Africa) and *Tragulus* (Asia): both are brush dwellers inhabiting primeval forests. The original group of the Pecora are **Hypertragulidae** (Eocene–Miocene, Asia, N. America).

2. Superfamily Cervoidea

Family Cervidae (deer) (figures 158, g, 161, a–i)

Hare- to horse-size Pecora, males are usually with antlers that, with increasing age, become stronger, larger, and more branched and are shed annually after the rut. Dental formula 0.1–0.3.3 / 3.1.3.3. Cannon bones. Metacarpals II and V usually incompletely developed;

320

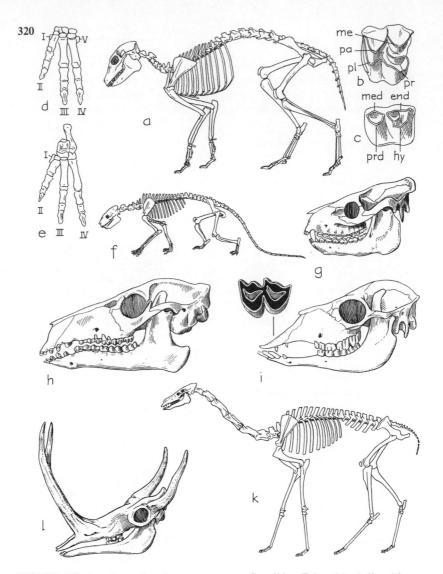

FIGURE 160. Ancodonta, Oreodonta, Tylopoda, Pecora. **a–c** *Cainotherium* (Oligocene, Europe, Cainotheriidae), **a** Skeleton, ≈30 cm long, **b** Upper M, **c** Lower M. **d, e** *Diplobune* (Eocene–Miocene, Europe, Anoplotheriidae), **d** Manus, **e** Pes. **f** *Agriochoerus* (Oligocene–Miocene, N. America, Agriochoeridae), skeleton, ≈2 m long. **g** *Oreodon* (Oligocene, N. America, Oreodontidae), skull, ≈13 cm. **h** *Poëbrotherium* (Oligocene, N. America, Camelidae, Tylopoda), skull, ≈16 cm. **i** *Lama* (Recent, guanaco, S. America, Camelidae, Tylopoda), skull and M, length of the skull ≈25 cm. **k** *Oxydactylus* (Miocene, N. America, Camelidae, Tylopoda), skeleton, ≈2.2 m long. **l** *Synthetoceras* (Miocene–Pliocene, N. America, Protoceratidae, Pecora), skull, ≈45 cm. Abbreviations as in figure **134** (after *Hürzeler, Leidy, Peterson, Schlosser, Scott, Stirton*).

sometimes they are retained only proximally (Plesiometacarpi), some-times only the distal parts (Telemetacarpi) remain. Cervids are divided into **Moschinae** (primitive; *Moschus,* recent, Asia), **Hydro-potinae** (*Hydropotes,* Recent, China), and the form-rich **Cervinae.** Cervinae emerged as antlerless forms during the European Oligocene–Miocene (*Amphitragus*) and North American Miocene (*Blastomeryx*). The most primitive living antlered deer are the **Muntjacinae** of South Asia, with small spiked or forked antlers. This stage of antler development had been reached during the Miocene (*Dicroceros*). The size and variety of antlers increased in several lines of these plesiometecarpal deer. The genus *Cervus* appeared in the Upper Pliocene. A culmination of antler development is found in the Pleistocene giant stags (*Megaceros*). The origin of the roe deer (*Capreolus*) is unclear. Telemetacarpal New World deer (**Odoco-ilinae**) are represented by numerous species that reached South America in the Pleistocene. The phylogeny of reindeer (**Rangiferi**) is still unknown.

3. Superfamily Giraffoidea (figures 158, m, n, 161, k–m)

Pecora of moderate to considerable size, whose skull bears (usually life-long) bony protuberances that are covered with skin in living spe-cies. Dental formula 3–0.1–0.4–3.3 / 3.1.4–3.3, lower canines bilo-bate. Giraffoidea have common ancestors with the Cervoidea. Three families can be distinguished: New World **Dromomerycidae** (Mio-cene–Pliocene), Old World **Palaeomerycidae** (Miocene–Pliocene), and **Giraffidae.** Giraffes are medium to very large Pecora with paired protuberances. Dental formula 0.0.3.3 / 3.1.3.3. Primitive representa-tives (short-necked giraffes) emerged during the North African Lower Miocene (*Palaeotragus*) and reached Eurasia during the Upper Mio-cene. Their last representative (*Okapia*) is a secondary forest dweller, occurring in the Congo. Long-necked giraffes (Giraffinae) appeared in the Upper Miocene and soon became widely distributed throughout Africa and Eurasia. In the post-Pleistocene they retreated to regions south of the Sahara. The large, short-necked Sivatheriinae (*Sivathe-rium,* Upper Miocene–Pleistocene, S. Asia, Africa) represent a sep-arate line.

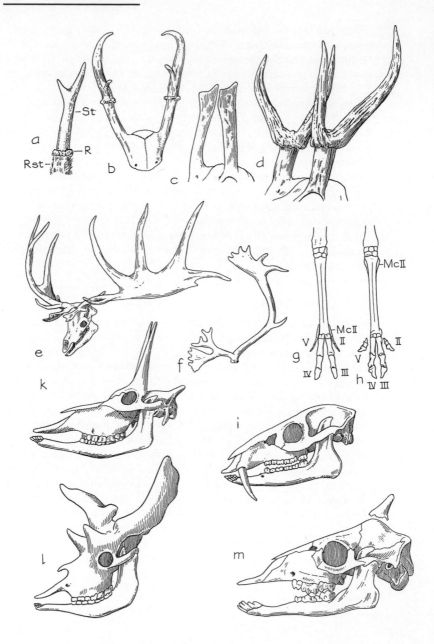

4. Superfamily Antilocaproidea

Family Antilocapridae (pronghorns) (figure 162, a, c)

A form-rich group of New World Pecora (Miocene–Recent). The only
surviving species is *Antilocapra americana,* with forked horns that
are annually shed and regrown.

5. Superfamily Bovoidea

Family Bovidae (Cavicornia, bovids) (figures 158, f, s, 162, b, d–f)

The most varied and species-rich group of the Artiodactyla, including
both small and large representatives. Their stretched-out frontal bones
carry bony processes (in both genders or only in males) with keratinous sheaths. Dental formula 0.1–0.2–3.3 / 3. 1.2–3.3, lower
canine is incisiviform. Cannon bones with sharp distal keels. Distribution: Lower Eocene–Recent. Flourished during Pliocene–
Pleistocene. Their homeland is Eurasia: a few types migrated to North
America. Africa, their current main stronghold, was colonized late. It
is assumed that bovids arose from Gelocidae (Traguloidea) of the
Oligocene. The oldest bovid known to date is the small *Eotragus*
(Lower Miocene, Eurasia, Africa), with short, nearly straight horny
pegs. Bovid classification presents great difficulties. Only a little
information can be gleaned from their great variety of forms.

Cephalophinae (duikers): Small, primitive brush dwellers
(Pleistocene–Recent, Africa).

Bovinae (bovines): Wild oxen appeared in the European Pliocene
with *Parabos.* Aurochs (*Bos primigenius*) emerged during the Middle
Pleistocene in Eurasia and was extirpated in 1627. It is regarded as
the ancestral form of domestic cattle. The *Bison* group has been

FIGURE 161. Cervidae (deer). a–f Antlers.
a, b *Muntjacus* (Recent, muntjac,
Palaeomerycidae); R, burr, Rst, persisting pedicle, St, beam; c, d *Dicroceros* (Miocene, Cervidae), c Juvenile
stage, d Full-grown antlers; e *Megaceros* (Pleistocene, giant stag, Cervidae); f *Rangifer caribou* (Recent,
reindeer, Cervidae). g *Blastomeryx*
(Miocene–Pliocene, Palaeomerycidae),
telemetacarpal manus. h *Cervus*
(Recent, Cervidae), plesiometacarpal
manus. i *Blastomeryx,* skull, ≈35 cm.
k *Samotherium* (Miocene–Pliocene,
Giraffidae), skull, ≈60 cm. l *Sivathe-
rium* (Pleistocene, Giraffidae), skull,
≈40 cm long. m *Giraffa* (Recent,
giraffe, Giraffidae), skull, ≈50 cm long
(after *Bohlin, Colbert, Forsyth Major,
Matthew, Schaub, Scott, Stehlin*).

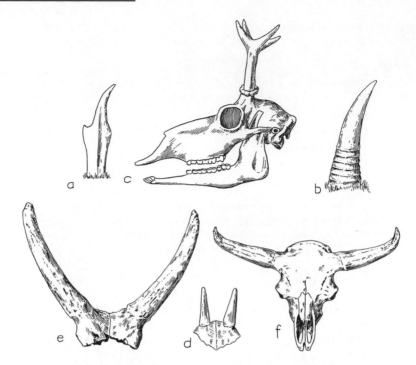

FIGURE 162. Bovoidea. **a** *Antilocapra* (Recent, pronghorn antelope, N. America, Antilocapridae), bony core, the keratinous sheath is shed yearly. **b** *Bos* (Recent, domestic cow), bony core, the horny sheath persists, at the basis are discernible the growth zones. **c** *Merycodus* (Miocene–Pliocene, N. America, Antilocapridae), skull, ≈18 cm. **d** *Eotragus* (Miocene, Africa, Asia, Europe, Bovidae). **e** *Parabos* (Pliocene, Europe, Africa, Bovidae). **f** *Bison priscus* (Pleistocene, Europe, Bovidae) (after *Frick, Stehlin*).

known since the beginning of the Pleistocene. The "buffalo" (bison; *Bison bison*) is a latecomer to North America. Bovid characteristics are most highly developed in bovines.

Tragelaphinae (bushbucks): Widely distributed in Africa since the Pliocene–Pleistocene.

Hippotraginae (hippotragine antelopes): Since the Upper Miocene in Eurasia, recent distribution in Asia (Near East) and Africa.

Reduncinae (water- and reedbuck): During the Pliocene and older Quaternary in South Asia, today living in Africa.

Alcelaphinae (hartebeest): During the Upper Miocene in North Africa, today in Africa and Asia (Near East).

Antilopinae (antelopes and gazelles): Many species. During the younger Tertiary they were distributed in Eurasia and North Africa. Recent gazelles live mainly in Africa.

Caprinae (goats and sheep): Appeared in Eurasia during the Middle Miocene. Differentiation at the end of the Tertiary and during the Pleistocene. *Capra aegagrus* (Asia) is the ancestral form of domestic goats. Wild sheep (*Ovis*) reached North America from Eurasia. Domestic sheep derived from *Ovis ammon* (Asia).

Ovibovinae (muskoxen): *Ovibos moschatus* evolved from plains forms. At the end of the Pleistocene, they were at home in Europe and North America; today they occur in Greenland and Alaska.

Saiginae (saigas): *Saiga tatarica* from the eastern steppes temporarily occupied Europe during the younger Pleistocene.

Rupicaprinae (chamois): Ancient bovids, today with a disjunct distribution over the northern hemisphere. Chamois (*Rupicapra*) in the mountains from the Pyrenees to Asia Minor, mountain goat (*Oreamnos*) in the mountains from southern Alaska to southern Montana, goral (*Naemorhedus*) in the mountains from Kashmir to Amurland.

THE "SUBUNGULATES" (HYRACOIDEA, PROBOSCIDEA, SIRENIA)

In spite of their different adaptations, Hyracoidea (hyraxes, dassies), Proboscidea (elephants), and Sirenia (sea cows) can be grouped as "subungulates" on the basis of their common anatomical and biochemical characteristics. Their descendancy from Condylarthra is quite certain.

14. Order Hyracoidea (hyraxes, dassies) (figure 163, a–d)

Primitive, secondarily plantigrade subungulates of small or large stature. Originally full dentition (3.1.4.3 / 3.1.4.3) was later partly reduced and specialized. Macrosmatic. The innermost toes of both the front and hind feet carry grooming claws; other fingers and toes have nails. Originally long-legged runners, they later developed plantar pads. Hyracoidea developed the process of rumination independently of the "true ungulates." Distribution: Middle Eocene–Recent. Two main branches could already be distinguished in the Lower Oligocene.

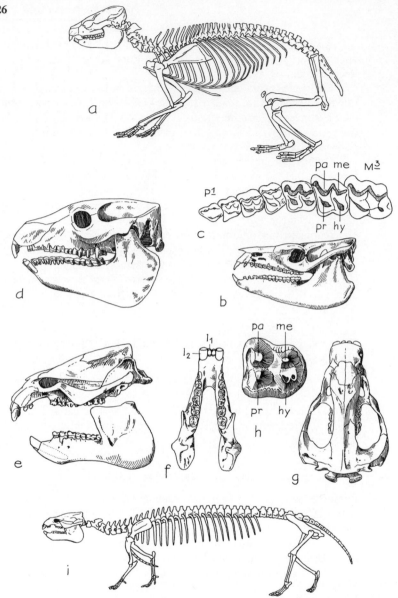

FIGURE 163. a–d Subungulata. **a** *Procavia* (Recent, hyrax, Hyracidae), skeleton, ≈50 cm long. **b, c** *Megalohyrax* (Oligocene, Geniohyidae), **b** Skull, ≈30 cm long, **c** upper P1–M3. **d** *Pliohyrax* (Pliocene, Hyracidae), skull, ≈24 cm long. **e–i** Proboscides, *Moeritherium* (Oligocene, Moeritheroidea), **e–g** Skull and lower jaw; skull ≈45 cm long, **h** upper M, **i** Restoration of the skeleton. Abbreviations as in figure **134** (after *Gregory, Hünermann, Osborn, Thenius, Young*).

1. Family Geniohyidae

Long-nosed, large Hyracoidea with a full dentition. Very large tri-dactylous forms occurred during the Miocene. Distribution: Lower Oligocene–Miocene.

2. Family Hyracidae

Short-faced forms with specialized front teeth. Dental formula 1.0. 4.3 / 2.0.3–4.3. Upper incisors are long, rootless; lower ones are chisel-shaped. A macroform is *Gigantohyrax* (Pleistocene, S. Africa). Distribution: Lower Oligocene–Recent. The three surviving genera (*Procavia, Heterohyrax, Dendrohyrax*) are small forms adapted to different habitats.

15. Order Proboscidea (elephants) (figures 163, e–i, 164, 165, a, b)

Mostly large to very large, pentadactylous, herbivorous subungulates. Their size requires far-reaching adaptations of the skeleton. The limbs are pillarlike, pes and manus short, digitigrade with plantal pads, mesaxonic, with hooflike nails. Various lines developed the trunk: enormously lengthened nose and upper lip evolved into a tactile and prehensile organ. The originally full dentition became highly specialized. Formation of the trunk and specialization of the dentition led to profound modifications of the skull: the nasal opening shifted backward, the jaw lengthened or shortened, and the skull became highly arched with large air cells in the diploe (inner layer of the skull bones). Distribution: Middle Eocene–Recent.

Proboscidea are of African origin; in the Upper Tertiary individual lines migrated to Eurasia, then pushed on over the Bering Strait to North America and reached South America during the Pleistocene. The only Recent representatives are *Loxodonta africana* in Africa and *Elephas maximus* in South Asia. Three suborders have been distinguished:

1. Suborder Moeritheroidea (figure 163, e–i)

Moderate-size Proboscidea with long body but no trunk. The only known genus, *Moeritherium*, led an amphibious life. It belongs to a sterile sideline. Distribution: Eocene–Lower Oligocene.

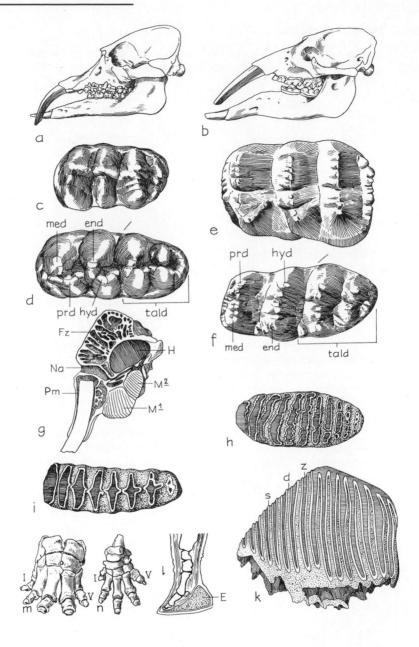

2. Suborder Elephantoidea (figure 164)

Elephantid group with the greatest number of species and forms. Dental formula 1.0.3–0.3 / 1–0.0.3–0.3, incisors developed into permanently growing tusks. Cheek teeth: cusps arrayed in parallel rows, brachyodont or hypsodont. The number of simultaneously functioning cheek teeth is variable. Bunodont and zygodont molars can be derived from a tribosphenic pattern. Bunodont molars have cross-girders of knoll-shaped elements; in the zygodont molars, the girders are developed into rather sharp ridges. The structure of molars and the type of lower jaw are taxonomically important. Distribution: Eocene–Recent.

1. Family Gomphotheriidae (Mastodontidae)

Oldest form is *Phiomia* (Lower Oligocene, Africa). Dental formula 1.0.3.3 / 1.0.2.3, all the bunodont molars simultaneously functional. Incisors developed as tusks. Their direct descendants are platybelodonts, with spadelike lower incisors (Miocene–Pliocene). *Gomphotherium (Mastodon)*, another descendant, has long lower tusks. It was numerous in Eurasia during the Miocene and in North America at the turn of the Miocene/Pliocene. The lower tusks disappear with the shortening of the lower jaw (elephantoid lower jaw). Further differentiation took place with the transition from trilophodont to tetralophodont teeth. Mastodons reached South America during the Pleistocene; by the beginning of the Holocene they were extinct.

2. Family Elephantidae

Descendants of mastodons, with molars composed of numerous transverse lamellae and with horizontal tooth replacement. Only two cheek

FIGURE 164. Proboscidea. **a** *Phiomia* (Oligocene, Gomphotheriidae), skull, ≈70 cm. **b** *Gomphotherium* (Lower Pliocene, Gomphotheriidae), skull, ≈80 cm. **c–f** Cheek teeth of the Proboscidea, **c, d** Bunodont, **e, f** Lophodont. **g** *Elephas maximus* (Recent), sagittal section of the skull; Fz, pneumatization of the frontal and parietal bones, H, cranial cavity, Na, nasal passage, Pm, premaxilla. **h** *Elephas maximus,* occlusal surface of an upper M. **i** *Loxodonta africana,* occlusal surface of an upper M. **k** *Elephas maximus,* sagittal section of an upper M; d, dentine (dotted), s, enamel lamella, z, cementum (hatched). **l–n** *Elephas maximus,* l Longitudinal section of manus; E, elastic pad, m, manus, n, pes, further abbreviations as in figure **134** (after *Marsh, Osborn, Weber*).

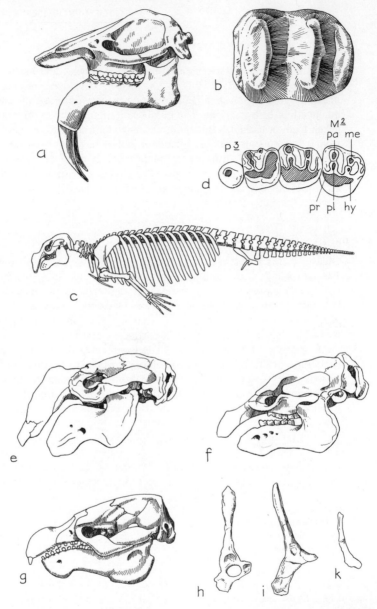

FIGURE 165. a, b Proboscidea. *Dinotherium* (Miocene, Dinotheroidea), **a** Skull, ≈1.4 m, **b** Upper M. **c–k** Sirenia. **c** *Halitherium* (Oligocene, Dugongidae), skeleton, ≈3 m long. **d** *Miosiren* (Miocene, Dugongidae), upper P3–M2; me, metalophe, pl, protolophe. **e** *Dugong* (Recent), skull, ≈30 cm. **f** *Manatus*, (Recent, *Trichechus*, Manatidae), skull, ≈28 cm. **g** *Protosiren* (Eocene, Dugongidae), skull, ≈30 cm. **h–k** Sirenian pelvis, **h** *Protosiren*, **i** *Halitherium*, **k** *Dugong*. Further abbreviations as in figure **134** (after *Abel, Deperet, Gaudry, Thenius*).

teeth function simultaneously. Three main lines can be derived from *Primelephas* (Upper Miocene, E. Africa): *Mammuthus, Loxodonta,* and *Elephas*. Mammoths were inhabitants of cold steppes during the youngest Pleistocene. The recent *Loxodonta africana* derives from *Loxodonta aurora* (Middle Pliocene, Africa). *Elephas* had several branches, among them Eurasian *Elephas antiquus* (wood elephant). From that form, *Elephas maximus* developed in Asia, and dwarf forms developed in the Mediterranean islands.

3. Family Mammutidae

Group of zygodont forms that separated during the Miocene from the bunodont mastodons. Characteristic genera are *Zygolophodon* (Miocene) and *Mammut* (not to be confused with *Mammuthus*) of the Pliocene–Pleistocene. Lower tusks were reduced as well. They survived in North America until the Holocene. Further descendants are **Stegodontidae** (Pleistocene, Asia, Africa), with tropical and subtropical forms.

3. Suborder Dinotheria (figure 165, a, b)

Fossil, medium to very large proboscids, with a pair of powerful, upward turned tusks and a trunk. Dental formula 0.0.2.3 / 1.0.2.3. Cheek teeth brachyodont, lophodont. Normal tooth replacement. Dinotheria appeared in Africa during the Miocene, reached Eurasia, and survived in Africa until the early Pleistocene. Marked increase in size (*Dinotherium gigantissimum,* Pliocene, height at withers up to 4 m). The roots of Dinotheria are sought among the Barytheria of the North African Eocene.

16. Order Sirenia (sea cows) (figure 165, c–k)

Medium to very large, aquatic, herbivorous subungulates. Forelimbs were modified into fins, hindlimbs were reduced. Body is torpedo-shaped, with a horizontal tail fin. Massive skull. Nasal opening shifted backward. Dentition, originally full, soon became reduced and specialized. Cheek teeth lophodont or bunolophodont. Microsmatic. Most of the bones, notably the vertebrae and ribs, are massive and heavy (pachyostosis).

Recent Sirenia are represented by the marine dugong (*Dugong*) and in tropical coastal waters and rivers by the manatee, *Manatus (Trichechus)*. Dugong has only one pair of upper incisors, whereas manatee has no front teeth. In both species, a hard, horny plate occurs in place of the incisors. The oldest finds (**Prorastomidae**) are from the Middle Eocene. **Dugongidae (Halicoridae)** and **Trichechidae (Manatidae)** evolved in the Oligocene. The history of the Dugongidae is well known, because they were widely distributed during the Oligocene and Miocene. *Halitherium* (Oligocene–Miocene), with large upper incisors and a full cheek dentition, is a characteristic representative. Toothless Steller's sea cow (*Rhytina,* up to 6 m long) was extirpated in historic times.

17. Order Cetacea (whales) (figure 166)

Exclusively aquatic, nearly completely naked Eutheria with a streamlined body, short neck, and horizontal tail fin. Premaxillae and maxillae are beaklike, elongated. Outer nasal opening is shifted backward. Micro- or anosmatic. Main sense is hearing. Whales use ultrasound both for orientation and communication. Tympanicum and petrosal are generally fused into one massive bone element (cetoliths). Dentition is originally full, heterodont (3.1.4.3 / 3.1.4.3), later heavily modified or even entirely reduced. Fin-shaped extremities are pentadactylous or tetradactylous, without nails. Outer hind limbs absent, though present during embryogenesis. Whales derive from primitive Condylarthra (Mesonychidae). Distribution: Eocene–Recent.

Three suborders are distinguished within the order:

1. Suborder Archaeoceti (archaeocetes, zeuglodons)

Primitive whales that were already fully adapted to aquatic life. Dentition full, heterodont, diphyodont. Three families: **Protocetidae, Dorudontidae,** and **Basilosauridae.** Distribution: Eocene–Miocene.

FIGURE 166. Cetacea. **a–c** *Basilosaurus* (Lower Eocene, Archaeoceti). **a** Restoration of the skeleton, ≈17 m long. **b, c** Lateral and dorsal views of the skull. **d, e** *Prosqualodon* (Miocene, Odontoceti), lateral and dorsal views of the skull, ≈45 cm long. **f** Skeleton of a baleen whale (Mysticeti); Recent baleen whales may reach 6- to 30-m length. **g** Cross section of the head of a recent baleen whale (Recent, *Balaenoptera*); F, baleen, l, tongue. **h** *Delphius,* (Recent, Delphinoidea), skull, ≈60 cm long, homodont dentition. **i** Streamlined shape of a dolphin (Recent, *Tursiops,* Delphinoidea). **k** *Globicephalus* (Recent, blackfish, Delphinoidea), forelimb. **l** Cetolith (tympano-perioticum). Further abbreviations as in figure **127** (after *Abel, Andrews, Delage, Flower, Gregory, Starck, Weber*).

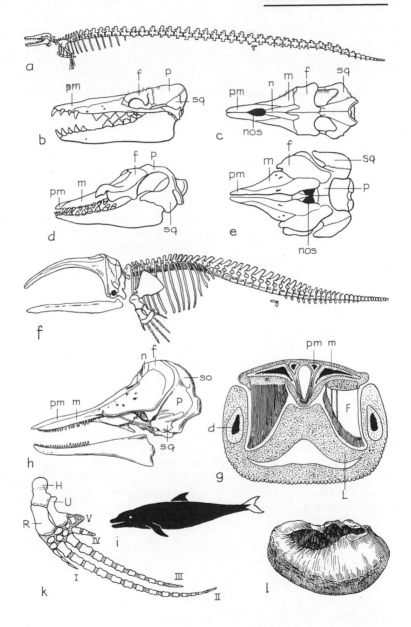

2. Suborder Mysticeti (baleen whales)

Largest of the living mammals (up to 30 m long), plankton eaters. Instead of teeth (seen only in the embryo), they possess baleen— keratinous plates suspended from the palate. Four families: **Cetotheriidae** (Oligocene–Pliocene), **Eschrichtiidae** (gray whales), **Balaenidae** (right whales), and **Balaenopteridae** (rorquals and humpback whales). Distribution: Oligocene–Recent.

3. Suborder Odontoceti (toothed whales)

Cetacean group containing the greatest number of species and forms. Distribution: Upper Eocene–Recent. The dentition of the oldest forms (**Agorophiidae**, Upper Eocene) is still complete. The teleoscopy of the skull (shortening and lateral displacement of parietal by supraoccipital and postparietal) had started. **Squalodontidae** (Oligocene– Miocene) exhibit an increase in the number of premolars. Primitive Recent representatives are river dolphins (**Platanistidae, Iniidae**). **Ziphiidae** (beaked whales, Miocene–Recent) have only two large teeth in the lower jaw. Among the **Physeteridae** (sperm whales, Miocene–Recent), the sperm whale or cachalot (up to 20 m long) has up to 60 homodont teeth, only in the lower jaw. **Delphinoidea** (dolphins and killer whales, Miocene–Recent), with numerous teeth in both jaws, are counted among the most highly evolved Odontoceti. The origin of **Monodontidae** (narwhal and beluga) is unknown.

18. Order Desmostylia (figure 167)

Marine amphibious, quadrupedal, herbivorous, fossil ungulates that reached almost the size of a hippopotamus. Cheek teeth are pillarlike with thick enamel, connected by cementum. Limbs functionally tetradactylous. Coastal forms of the Northern Pacific. Descendants of ancient hoofed mammals. Distribution: Oligocene–Miocene. Two families:

1. Family Cornwallidae

Dental formula 3.1.3.3 / 3.1.3.3, cheek teeth brachyodont. Distribution: Oligocene–Miocene.

2. Family Desmostylidae

Dental formula 0.1.3.3 / 1.1.3.3, cheek teeth hypsodont. Upper canines and lower incisors and canines developed into tusks. Distribution: Miocene.

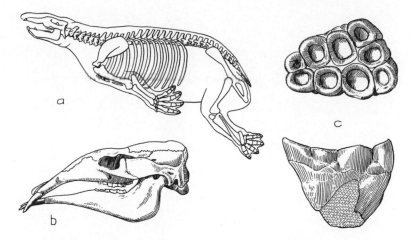

FIGURE 167. Desmostylia. **a** *Palae-oparadoxa* (Miocene, Cornwalliidae). Skeleton with an outline of the body (swimming), ≈2.3 m long. **b, c** *Des-mostylus* (Miocene, Desmostylidae). **b** Skull, ≈0.8 m long, **c** Upper M, occlusal surface and lateral view (after *Gregory, Vanderhoof*).

19. Order Embrithopoda (figure 168)

Giant, herbivorous ungulates of which to date only the genus *Arsinoitherium* (up to 3 m) from the Egyptian Lower Oligocene is well known. Limbs pillar-shaped, pentadactylous, skull with two large nasal and two small frontal horns. Full dentition, molars hypsodont, bilophodont. Embrithopoda probably evolved early from primitive Condylarthra.

20. Order Notoungulata (figure 169)

Fossil group of long-lived, herbivorous South American ungulates, containing many forms. The same characteristic structure of the otic region is found in all of them. Phylogenic changes are manifested mainly in the dentition and the extremities. Dentition full or specialized (enlarged incisors, fang-shaped canines). Teeth originally brachyodont, may become hypsodont with roots or rootless, prismatic. Lophodont cheek teeth are dominant. Limbs are usually specialized, with two to five functional metapodia. Feet mainly mesaxonic, rarely paraxonic, plantigrade or semiplantigrade, frequently with hooves, seldom with large or small claws. Notoungulata are considered to be descendants of the Paleocene Condylarthra. Dis-

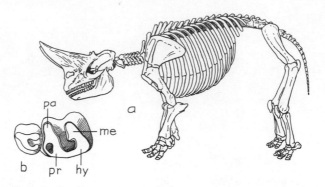

FIGURE 168. Embrithopoda. **a, b** *Arsinoithe-rium* (Lower Oligocene, Arsinoitherii-dae). **a** Restoration of the skeleton, ≈3.3 m long, **b** Upper P4, M1. Abbre-viations as in figure **134** (after *Andrews*).

tribution: Paleocene–Pleistocene. Notoungulata encompass four suborders:

1. Suborder Notioprogonia (figure 169, a, e)

   Primitive, relatively small forms, little specialization. Distribution: Paleocene–Eocene. They probably lived both in Asia (Paleocene) and North America (Lower Eocene).

2. Suborder Toxodonta (figure 169, b, f–k)

   The largest group, with tridactylous, mesaxonic extremities. Flour-ished during the Miocene (*Homalodotherium*). Large representatives survived into the Pleistocene (*Toxodon*). Distribution: Paleocene–Pleistocene.

3. Suborder Typotheria (figure 169, c, l, m)

   There is a trend to reduce the lateral toes of both the front and hind limbs. Distribution: Eocene–Pliocene. *Miocochilius* (Miocene) had didactylous, paraxonic extremities.

4. Suborder Hegetotheria (figure 169, d)

   Skeleton and dentition resemble that of a hare (*Hegetotherium*, Mio-cene). Distribution: Eocene–Pleistocene.

21. Order Pyrotheria

Fossil, large, herbivorous ungulates of South America. Only the genus *Pyrotherium* (Oligocene) is well known; it resembles the elephant. Nasal opening shifted backward (trunk). Dental formula 2.0.3.3 / 1.0.2.3, one pair of both upper and lower incisors modified into tusks. Cheek teeth lophodont. Extremities are pillar-shaped, digitigrade, front limbs shorter than hind ones. Manus and pes pentadactylous. Pyrotheria probably derived from Condylarthra of the Paleocene. Distribution: Upper Eocene–Lower Oligocene.

22. Order Astrapotheria

Medium to very large, amphibious. herbivorous ungulates from South America. Short nasals as well as location of the nasal opening on the upper side of the skull indicate the presence of a trunk. Dental formula 0.1.3–1.3 / 3.1.2–1.3, canines grow into tusks. Molars are hypsodont, lophodont. Extremities pentadactylous with hooves. Manus is digitigrade, with plantar pads, pes plantigrade. Origin unknown. Distribution: Paleocene–Miocene.

   Trigonostylopidae are an aberrant branch of South American ungulates that exhibit certain Astrapotherian as well as some Litopternian characteristics. Best known is *Trigonostylops* (Paleocene–Eocene), with full dentition. Distribution: Paleocene-Oligocene.

23. Order Tillodontia

Fossil, short-lived Eutheria of the Paleocene and Eocene. Dental formula 2.1.3.3 / 3–1.1.3.3. During the phylogeny, both upper and lower I 2 become hypsodont and rootless. Upper molars are trituberculate, lower lophodont. Omnivorous or herbivorous. Clavicles have been documented. Pentadactylous, clawed, plantigrade extremities. There was an increase in size during the evolution. *Tillodon* (Middle Eocene, N. America) reached the size of a brown bear. Tillodontian ancestors are sought among Arctocyonidae (Condylarthra).

24. Order Litopterna (figure 170)

Fossil South American mesaxonic ungulates of medium to considerable size. Dentition originally full, later reduced. Cheek teeth brachyodont or hypsodont. Upper molars lophodont or selenodont, lower are biselenodont. Their tendency toward tridactyly or mono-

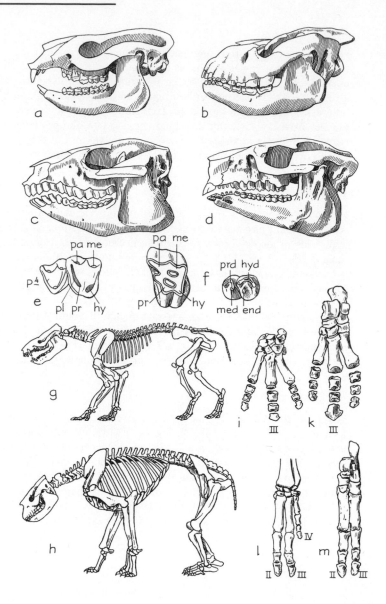

dactyly, such as is found in the equids, provides a classical example of convergent evolution. Distribution: Paleocene–Pleistocene.

Litopterna were the "horses" and "camels" of South America, with their respective families, Prototheriidae and Macraucheniidae. The two families separated as early as during the Upper Paleocene. The group derives from Condylarthra.

1. Family Prototheriidae

Forms of moderate size that, during the Oligocene/Miocene, achieved functional monodactyly. In the graceful *Thoatherium* (Miocene), reduction of the lateral toes progressed even further than in modern horses. Dental formula 1.0.4.3 / 2.1.4.3.

2. Family Macraucheniidae

Representatives of this family had a trunk and achieved a functional tridactyly. The highest specialized form was *Macrauchenia* (Pliocene–Pleistocene), which stood 1.8 m high at the withers.

25. Order Pantodonta (figure 171, a–d, f, g)

Fossil, archiac Eutheria of the early Tertiary (N. America, Eurasia), medium to very large. Dentition originally full. Extremities pentadactylous, terminal phalanges nail- to hoof-shaped. Early semi-digitigrade, later digitigrade forelimbs, hindlimbs plantigrade. Distribution: Paleocene–Middle Oligocene.

Two main branches are distinguished; they can be derived from Arctocyonidae (Condylarthra).

FIGURE 169. Notoungulata. **a** *Notostylops* (Eocene, S. America, Notioprogonia), skull, ≈14 cm long. **b** *Homalodotherium* (Miocene, S. America, Toxodontia), skull, ≈38 cm. **c** *Protypotherium* (Miocene, S. America, Typotheroidea), skull, ≈10 cm. **d** *Hegetotherium* (Miocene, S. America, Hegetotheroidea), skull, ≈11 cm. **e** *Henricosbornia* (Upper Paleocene, S. America, Notioprogonia), upper P4, M1. **f** *Oldfieldthomasia* (Eocene, S. America, Toxodontia), upper and lower M. **g** *Thomashuxleya* (Eocene, S. America, Toxodontia), skeleton, ≈1.5 m long. **h** *Homalodotherium* (Miocene, S. America, Toxodontia), skeleton, ≈1.8 m long. **i, k** *Nesodon* (Miocene, S. America, Toxodontia), **i** Manus, **k** Pes. **l, m** *Miocochilius* (Upper Miocene, S. America, Typotheroidea), **l** Manus, **m** Pes. Abbreviations as in figure 134 (after *Gregory, Riggs, Schlosser, Scott, Simpson, Sinclair, Stirton*).

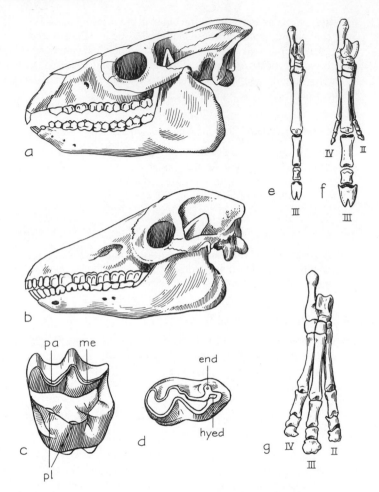

FIGURE 170. Litopterna. **a** *Thoatherium* (Lower Miocene, S. America, Proterotheriidae), skull, ≈18 cm long. **b** *Macrauchenia* (Pleistocene, S. America, Macraucheniidae), skull, ≈45 cm long. **c, d** *Proterotherium* (Lower Miocene, S. America, Protero- theriidae), **c** Upper M, **d** Lower M. **e** *Thoatherium,* pes. **f** *Diadiaphorus* (Lower Miocene, S. America, Proterotheriidae), pes. **g** *Macrauchenia,* pes. Abbreviations as in figure **134** (after *Burmeister, Scott*).

1. Superfamily Pantolambdoidea

Relatively small, slim forms. Distribution: Paleocene.

2. Superfamily Coryphodontoidea

Distribution: Paleocene–Middle Oligocene. Large forms; *Coryphodon* (Upper Paleocene–Lower Eocene) was nearly 3 m long. Dentition suggests omnivorous-herbivorous food habits.

26. Order Dinocerata (figure 171, e, h, i)

Fossil, large, omnivorous or herbivorous Eutheria of Asian and North American older Tertiary. Horn-shaped structures may be developed on the skull. Dentition originally full; later the incisors were reduced and the canines were either enlarged or reduced. Distribution: Paleocene–Eocene.

Dinocerata are a parallel group of the Pantodonta. Three families are distinguished, derived from Arctocyonidae (Condylarthra).

1. Family Prodinoceratidae

These oldest representatives reached bear size. Distribution: Upper Paleocene–Lower Eocene.

2. Family Uintatheriidae

Herbivorous forms carrying hornlike structures. Distribution: Eocene; period of greatest prosperity was the late Eocene. *Uintatherium*, dental formula 0.1.3.3 / 3.1.3.3, upper canine large, saber-shaped.

3. Family Gobiotheriidae

Gobiotherium, the size of a rhinoceros, semiaquatic. Dental formula 0.0.3.3 / 3.1.3.3. Distribution: Middle Eocene.

27. Order Pholidota (pangolins, scaly anteaters) (figure 172)

Family Manidae

Unguiculate, plantigrade Eutheria whose tail and upper side of the body are covered with overlapping horny scales, with sparse hair between them. Extremities are pentadactylous or tetradactylous, ter-

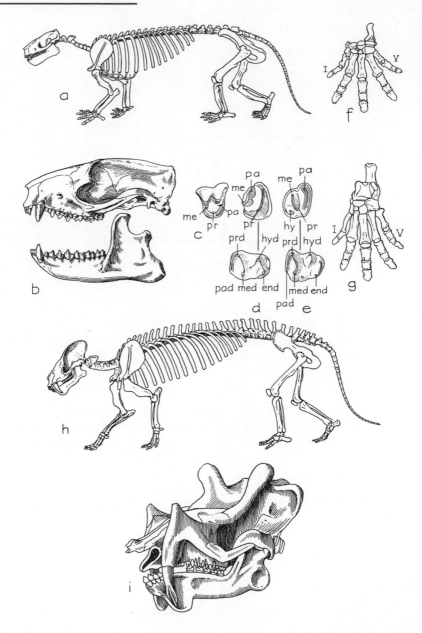

minal phalanges are split and carry strong claws. Toothless jaws, long, sticky tongue (myrmecophagous). Distribution: Eocene–Recent.

Phylogeny and relationships are still completely unknown. The single recent genus lives in the tropics of Asia and Africa. *Eomanis* (Middle Eocene, Europe) already showed the same type of biological adaptations as do the recent representatives.

Metacheiromyidae (Middle Eocene, N. America) are also regarded as members of the Pholidota (figure 172, d).

28. Order Tubulidentata (tubulidentates) (figure 173)

Family Orycteropidae (aardvark)

Unguiculate, insectivorous Eutheria, currently represented by only one species (Africa south of the Sahara). Skull with a long, tubular snout. Macrosmatic. Dental formula 0.0.2. 3 / 0.0.2.2–3. Homodont cheek teeth without enamel, hypsodont, rootless, with a coat of cement (ant and termite eater). Forelimbs tetradactylous, hindlimbs pentadactylous with strong claws. Distribution: Miocene–Recent.

Tubulidentata probably represent an old African group. The postcranial skeleton suggests origins in the primitive Condylarthra. The oldest finds are from the African Lower Miocene; in the Middle Miocene the animals also reached Eurasia.

29. Order Taeniodonta ("Ganodonta")

Isolated group of fossil Eutheria, known from early Tertiary, primarily from North America. Only one family, **Stylodontidae** (Paleocene–Eocene), with two separate branches; origins not yet clarified.

30. Order Xenarthra (edentates) (figures 174, 175)

Small to very large terrestrial or arboreal Eutheria of South America. Insectivorous, omnivorous, or herbivorous. Dense or sparse hair, fre-

---

FIGURE 171. Pantodonta and Dinocerata. **a, b** *Pantolambda* (Paleocene, Pantolambdoidea), **a** Skeleton, ≈1 m long, **b** Skull, ≈16 cm long. **c–e** Molars. **c** *Pantolambda,* **d** *Coryphodon* (Lower Eocene, N. America, Europe, Coryphodontoidea). **e** *Uintatherium* (Upper Eocene, N. America, Uintatheroidea). **f, g** *Pantolambda,* **f** Manus, **g** Pes. **h** *Mongolotherium* (Upper Eocene, Asia, Prodinoceratidae), skeleton, ≈3 m long. **i** *Uintatherium,* skull, ≈80 cm long. Abbreviations as in figure **134** (after *Marsh, Matthew, Osborn, Simons, Simpson*).

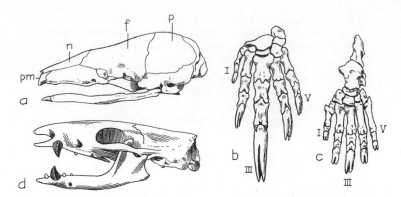

FIGURE 172. Pholidota. **a–c** *Manis* (Recent, pangolin, Manidae). **a** Skull, ≈9 cm long, **b** Manus, **c** Pes. **d** Metacheiromyidae, *Metacheiromys* (Middle Eocene, N. America), lateral view of the skull, approx. half life-size. Abbreviations as in figure **127** (after *Anthony, Grassé, Simpson*).

quently with horny scales and ossifications of the leathery skin that can form an armor. Thoracic and lumbar vertebrae have accessory joints (xenarthral articulation). Ischium is connected with sacrum by a joint. Manus pentadactylous, tetradactylous, or tridactylous, pes pentadactylous, plantigrade. Terminal phalanges generally have claws. Dentition often monophyodont, with four to ten teeth on each side, rarely more. Cheek teeth prismatic, predominantly hypsodont and rootless, without enamel, Myrmecophaga are toothless. Macrosmatic.

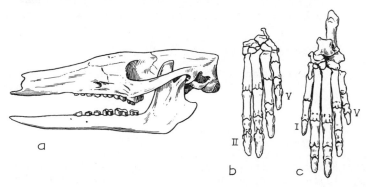

FIGURE 173. Tubulidentata. **a** *Orycteropus* (Pliocene, aardvark, Orycteropidae), skull, ≈16 cm long. **b, c** *Orycteropus* (Recent), **b** Manus, **c** Pes (after *Anthony, Colbert*).

Archaic characteristics occur (septomaxillas, low body temperature, imperfect thermoregulation). Distribution: Paleocene–Recent.

This characteristic South American group is divided into three suborders, of which the Pilosa and the Cingulata are supposed to have derived from primitive insectivores.

1. Suborder Pilosa (sloths) (figure 174, d–g)

Hairy arboreal or terrestrial forms with at most four or five teeth. Distribution: Oligocene–Recent. Recent representatives are phyllophagous **Bradypodidae** (three-toed sloths) that live in tropical forests. Related to them are the giant sloths, of which the oldest representatives are known from the Oligocene. Giant sloths arrived in North America during the Pleistocene and became extinct in the early Holocene.

2. Suborder Cingulata (armadillos) (figure 174, h–k, 175, a–c)

Burrowing xenarthrans whose head, body, and tail are protected by an armor of bony plates and horny scales. Distribution: Paleocene–Recent. Today they are represented only by **Dasypodidae,** of which several species are polyembryonic (four to twelve young of the same sex can grow from a single cell). The **Glyptodontidae** that produced the giant armadillos during the Pliocene separated from the dasypodids as early as during the older Tertiary. They reached North America during the Pleistocene and disappeared by the beginning of the Holocene.

3. Suborder Vermilingua (anteaters) (figure 175, d)

Distribution: Miocene–Recent. Recent representatives (**Myrmecophagidae**) have a small head with a tubular, toothless snout and a protrusible tongue (ant and termite eaters). Scarce fossil documents provide no clues about their origin.

Discovery of a xenarthran was recently reported in Paleocene deposits in China.

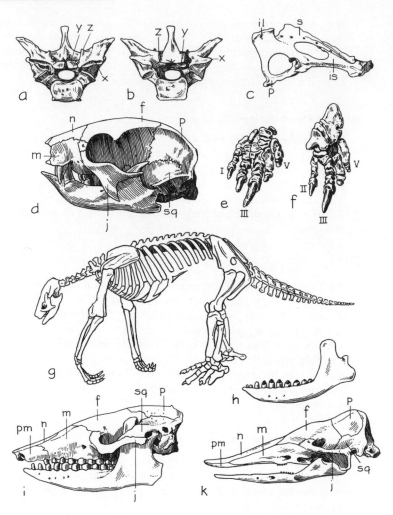

FIGURE 174. Xenarthra. **a, b** Lumbar vertebra, **a** Anterior, **b** Posterior view; x, y, z, xenarthral articulation. **c** Pelvis; il, ilium, is, ischium, p, pubis, s, sacrum. Note the synostosis of ischium and sacrum. **d** *Choloepus* (Recent, S. America, sloth with six cervical vertebrae), skull, ≈8.5 cm long. **e, f** *Mylodon* (Pleistocene, giant sloth, Mylodontidae), **e** Manus, **f** Pes. **g** *Nothrotherium* (Pleistocene, giant sloth, Megatheriidae), skeleton, ≈2.5 m long. **h** *Utaetus* (Lower Eocene, Dasypodidae), lower jaw, approx. half life-size. **i** *Euphractus* (Recent, Dasypodidae), skull, ≈10 cm long. **k** *Stegotherium* (Lower Miocene, Dasypodidae), skull, ≈12 cm long. Abbreviations as in figure **127** (after *Grassé, Hoffstetter, Owen, Scott, Simpson, Starck, Stock, Weber*).

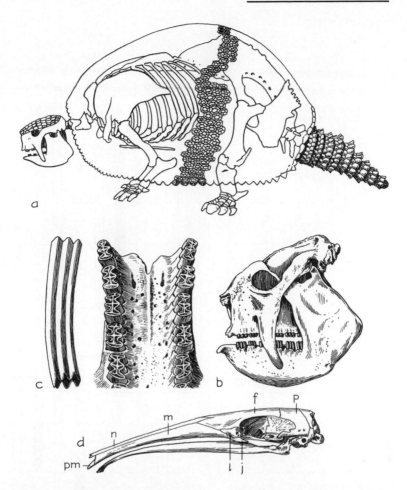

FIGURE 175. Xenarthra. **a, b** *Glyptodon* (Pleistocene, giant armadillo, Glyptodontidae). **a** Restoration of the skeleton, ≈3 m long, **b** Skull, ≈28 cm long. **c** *Panochthus* (Pleistocene, Glyptodontidae), dentition of the palate, and a lateral view of one molar. **d** *Myrmecophaga* (Recent, giant anteater, Myrmecophagidae), skull, ≈22 cm long. Abbreviations as in figure **127** (after *Burmeister, Scott, Weber*).

Overviews

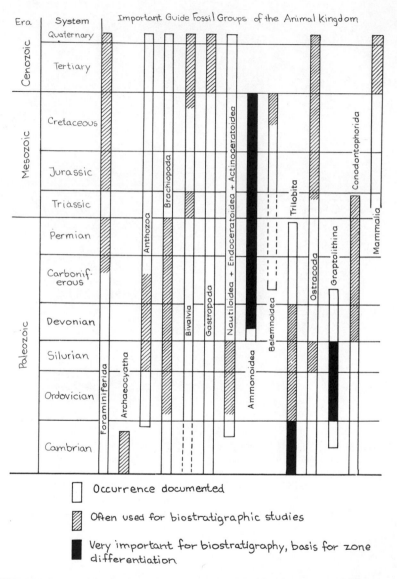

FIGURE 176. Temporal distribution (geological range) of the most important guide fossil groups.

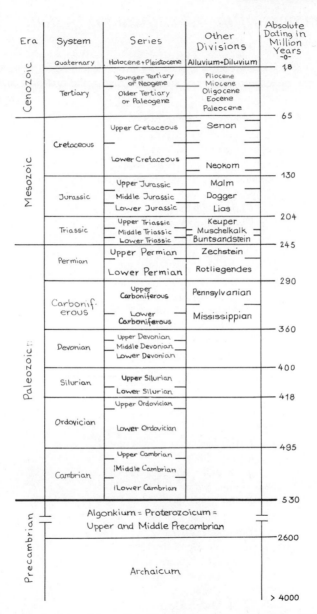

FIGURE 177. Geologic time and formations (absolute dating after G. S. Odin, 1982).

Summary of the Classification System

The list is not complete, because taxa without fossil representatives are included only if they are mentioned in the text or if they are important for continuity. The subdivision of individual groups varies.

1. Superkingdom: **Procaryota**
2. Superkingdom: **Eucaryota**
1. Kingdom: *Plantae (Plants)*
2. Kingdom: *Animalia (Animals)*
1. Subkingdom: Protozoa *(Monocellular Animals)*
   1. Phylum: Protozoa
      1. Class: Flagellata
      2. Class: Rhizopoda
         1. Order: Amoebida
         2. Order: Arcellinida
         3. Order: Foraminiferida
         1. Suborder: Allogromiina
         2. Suborder: Textulariina
         3. Suborder: Fusulinina
         4. Suborder: Miliolina
         5. Suborder: Rotaliina
         4. Order: Heliozoa
         5. Order: Radiolaria
         1. Suborder: Acantharia
         2. Suborder: Spumellina
         3. Suborder: Nassellina
         4. Suborder: Phaeodaria
      3. Class: Sporozoa
      4. Class: Ciliata
      Calpionellidea (Protista incertae sedis)
2. Subkingdom: Metazoa (Multicellular)
   1. Division: Mesozoa
   2. Division: Parazoa
      1. Phylum: Porifera
         1. Class: Demospongea
            1. Order: Keratosida
            2. Order: Haplosclerida
            3. Order: Peocilosclerida
            4. Order: Hadromerida
            5. Order: Epipolasida
            6. Order: Choristida

      7. Order: Carnosida
      8. Order: Lithistida
   2. Class: Hexactinellida
      1. Order: Lyssakida
      2. Order: Dictyida
      3. Order: Lychniskida
   3. Class: Calcispongea
      1. Order: Dialytina
      2. Order: Pharetronida
      1. Suborder: Inozoa
      2. Suborder: Sphinctozoa
      3. Order: Octactinellida
   4. Class: Sclerospongiae
      1. Order: Ceratoporellida
      2. Order: Tabulospongia
      3. Order: Merliida
      4. Order: Chaetetida
      5. Order: Stromatoporoidea
  2. Phylum: Archaeocyatha
3. Division: Eumetazoa
1. Subdivision: Coelenterata
  1. Phylum: Hydrozoa
    1. Class: Hydrozoa
    2. Class: Scyphozoa
    1. Subclass: Scyphomedusae
    2. Subclass: Conulata
    3. Class: Anthozoa
    1. Subclass: Cerianthipatharia
    2. Subclass: Octocorallia
      1. Order: Stolonifera
      2. Order: Telestacea
      3. Order: Alcyonacea
      4. Order: Trachypsammiacea
      5. Order: Helioporida
      6. Order: Gorgonacea
      7. Order: Pennatulacea
    3. Subclass: Zoantharia
      1. Order: Zoanthiniaria
      2. Order: Corallimorpharia
      3. Order: Actinaria
      4. Order: Rugosa
      5. Order: Scleractinia

    6. Order: Heterocorallia
    7. Order: Tabulata
   2. Phylum: Acnidaria

  2. Subdivision: Bilateria
   1. Series: Protostomia
    1. Phylum: Tentaculata
     1. Class: Phoronida
     2. Class: Bryozoa
     1. Subclass: Phylactolaemata
     2. Subclass: Stenolaemata
      1. Order: Cyclostomata
      2. Order: Trepostomata
      3. Order: Cryptostomata
     3. Subclass: Gymnolaemata
      1. Order: Cheilostomata
      2. Order: Ctenostomata
     3. Class: Brachiopoda
     1. Subclass: Inarticulata
      1. Order: Lingulida
      2. Order: Acrotretida
     2. Subclass: Articulata
      1. Order: Orthida
      2. Order: Strophomenida
      3. Order: Pentamerida
      4. Order: Rhynchonellida
      5. Order: Spiriferida
      6. Order: Terebratulida
    2. Phylum: Mollusca
    1. Subphylum: Amphineura
     1. Class: Solenogastres
     2. Class: Polyplacophora
    2. Subphylum: Conchifera
     1. Class: Monoplacophora
     2. Class: Gastropoda
     1. Subclass: Prosobranchia
      1. Order: Archaeogastropoda
      2. Order: Mesogastropoda
      3. Order: Neogastropoda
     2. Subclass: Opisthobranchia
      1. Order: Pleurocoelia
      2. Order: Pteropoda

3. Order: Sacoglossa
4. Order: Acoela
3. Subclass: Pulmonata
  1. Order: Basommatophora
  2. Order: Stylommatophora
3. Class: Scaphopoda
4. Class: Bivalvia
1. Subclass: Palaeotaxodonta
  1. Order: Nuculoida
2. Subclass: Cryptodonta
  1. Order: Solemyoidea
  2. Order: Praecardioida
3. Subclass: Pteriomorpha
  1. Order: Arcoida
  2. Order: Mytiloida
  3. Order: Pterioida
4. Subclass: Palaeoheterodonta
  1. Order: Modiomorphoida
  2. Order: Unionoida
  3. Order: Trigonioida
5. Subclass: Heterodonta
  1. Order: Veneroida
  2. Order: Myoida
  3. Order: Hippuritoida
6. Subclass: Anomalodesmata
  1. Order: Pholadomyoida
5. Class: Cephalopoda
1. Subclass: Nautiloidea
  1. Order: Ellesmerocerida
  2. Order: Orthocerida
  3. Order: Ascocerida
  4. Order: Oncocerida
  5. Order: Discosorida
  6. Order: Tarphycerida
  7. Order: Nautilida
2. Subclass: Endoceratoidea
3. Subclass: Actinoceratoidea
4. Subclass: Bactritoidea
5. Subclass: Ammonoidea
  1. Order: Anarcestida
  2. Order: Goniatitida
  3. Order: Prolecanitida

    4. Order: Clymeniida
    5. Order: Ceratitida
    6. Order: Ammonitida
    7. Order: Phylloceratida
    8. Order: Lytoceratida
    9. Order: Ancyloceratida
  6. Subclass: Coleoidea
    1. Order: Sepioidea
    2. Order: Teuthoidea
    3. Order: Octobrachia
    4. Order: Phragmoteuthida
    5. Order: Aulacocerida
    6. Order: Belemnitida

Mollusca incertae sedis
1. Class: Tentaculitoidea
2. Class: Calyptoptomatida
Group of phyla: Articulata (Phyla 3–5)
3. Phylum: Annelida
  1. Class: Polychaeta
  2. Class: Myzostomida
  3. Class: Clitellata
4. Phylum: Onychophora
5. Phylum: Arthropoda
1. Subphylum: Trilobitomorpha
  1. Class: Merostomoidea
  2. Class: Pseudonotostraca
  3. Class: Marrellomorpha
  4. Class: Trilobita
    1. Order: Agnostida
    2. Order: Redlichiida
    3. Order: Corynexochida
    4. Order: Ptychopariida
    5. Order: Phacopida
    6. Order: Lichida
    7. Order: Odontopleurida
2. Subphylum: Chelicerata
  1. Class: Merostomata
  1. Subclass: Xiphosura
  2. Subclass: Eurypterida
  2. Class: Arachnida
  3. Class: Pantopoda

3. Subphylum: Mandibulata
   1. Class: Crustacea with 9 Subclasses
      1. Subclass: Branchiopoda
      2. Subclass: Ostracoda
      3. Subclass: Cirripedia
      4. Subclass: Malacostraca
   2. Class: Myriapoda
   3. Class: Insecta
      1. Subclass: Apterygota with 5 orders
      2. Subclass: Pterygota with 14 superorders and 28 orders
         1. Superorder: Palaeodictyoptera
         2. Superorder: Odonatopteroida
2. Series: Deuterostomia
6. Phylum: Echinodermata
1. Subphylum: Homalozoa
   1. Class: Homostelea
   2. Class: Stylophora
   3. Class: Homoiostelea
2. Subphylum: Blastozoa
   1. Class: Eocrinoidea
   2. Class: Cystoidea
   3. Class: Blastoidea
3. Subphylum: Crinozoa
   1. Class: Paracrinoidea
   2. Class: Crinoidea
      1. Subclass: Camerata
      2. Subclass: Inadunata
      3. Subclass: Flexibilia
      4. Subclass: Articulata
4. Subphylum: Asterozoa
   1. Class: Stelleroidea
      1. Subclass: Somasteroidea
      2. Subclass: Asteroidea
      3. Subclass: Ophiuroidea
5. Subphylum: Echinozoa
   1. Class: Edrioasteroidea
   2. Class: Echinoidea
      1. Subclass: Perischoechinoidea with 4 orders
         1. Order: Cidaroidea
      2. Subclass: Euechinoidea with 4 superorders and 16 orders
   3. Class: Helicoplacoidea
   4. Class: Ophiocistioidea

   5. Class: Cyclocystoidea
   6. Class: Holothuroidea
Echinodermata incertae sedis
   1. Class: Camptostromoidea
   2. Class: Lepidocystoidea
7. Phylum: Hemichordata
   1. Class Enteropneusta
   2. Class: Pterobranchia
      1. Order: Rhapdopleurida
      2. Order: Cephalodiscida
   3. Class: Graptolithina
      1. Order: Dendroidea
      2. Order: Tuboidea
      3. Order: Camaroidea
      4. Order: Stolonoidea
      5. Order: Crustoidea
      6. Order: Graptoloidea

Conodonta (Phylum incertae sedis)
8. Phylum: Chordata
1. Subphylum: Urochordata
2. Subphylum: Cephalochordata
3. Subphylum: Vertebrata
   1. Superclass: Agnatha
   1. Class: Myxinomorpha
   2. Class: Cephalaspidomorpha
   3. Class: Pteraspidomorpha
   4. Class: Thelodonti
   2. Superclass: Gnathostomata
   1. Class: Placodermi
   2. Class: Chondrichthyes
   1. Subclass: Elasmobranchii
   2. Subclass: Subterbranchialia
   3. Class: Acanthodii
   4. Class: Osteichthyes
   1. Subclass: Actinopterygii
   1. Infraclass: Chondrostei
   2. Infraclass: Neopterygii
   2. Subclass: Sarcopterygii
   1. Infraclass: Dipnoi
   2. Infraclass: Crossopterygii
      1. Order: Actinistia
      2. Order: Holoptychiida

   3. Order: Osteolepididia
   4. Order: Rhizodontida
   5. Order: Onychodontida (Struniiformes)
5. Class: Amphibia
1. Subclass: Labyrinthodontia
   1. Order: Ichthyostegalia
   2. Order: Temnospondyli
   3. Order: Anthracosauria
2. Subclass: Lepospondyli
   1. Order: Aistopoda
   2. Order: Nectridia
   3. Order: Microsauria
3. Subclass: Lissamphibia
   1. Order: Anura
   2. Order: Urodela
   3. Order: Apoda
6. Class: Reptilia
1. Subclass: Anapsida
   Order: Protothyromorpha
   Order: Procolophonia
   Order: Pareiasauria
2. Subclass: Chelonomorpha
   Order: Testudines
3. Subclass: Eosuchiamorpha
   Order: Eosuchia
4. Subclass: Lepidosauria
   Order: Rhynchocephalia
   Order: Squamata
5. Subclass: Archosauria
   Order: Thecodontia
   Order: Crocodilia
   Order: Saurischia
   Order: Ornithischia
6. Subclass: Pterosauromorpha
   Order: Pterosauria
7. Subclass: Sauropterygomorpha
   Order: Sauropterygia
8. Subclass: Placodontomorpha
   Order: Placodontia
9. Subclass: Ichthyopterygia
   Order: Ichthyosauria

10. Subclass: Synapsida
　　Order: Pelycosauria
　　Order: Therapsida
7. Class: Aves
1. Subclass: Archaeornithes
　　Order: Archaeopterygiformes
2. Subclass: Neornithes
　　1. Superorder: Odontognathae
　　2. Superorder: Palaeognathae
　　3. Superorder: Neognathae
8. Class: Mammalia
1. Subclass: Prototheria s.l.
1. Infraclass: Eotheria
　　1. Order: Morganucodonta
　　2. Order: Docodonta
　　3. Order: Triconodonta
2. Infraclass: Prototheria s.s.
　　4. Order: Monotremata
3. Infraclass: Allotheria
　　5. Order: Multituberculata
2. Subclass: Theria s.l.
1. Infraclass: Trituberculata
　　1. Order: Kuehneotheria
　　2. Order: Symmetrodonta
2. Infraclass: Pantotheria
　　Order: Eupantotheria
3. Infraclass: Metatheria
　　Order: Marsupialia
4. Infraclass: Eutheria (Placentalia)
　　1. Order: Insectivora
　　2. Order: Macroscelidea
　　3. Order: Dermoptera
　　4. Order: Scandentia
　　5. Order: Chiroptera
　　6. Order: Primates
　　7. Order: Rodentia
　　8. Order: Hyaenodonta
　　9. Order: Carnivora
　　10. Order: Lagomorpha
　　11. Order: Condylarthra
　　12. Order: Perissodactyla
　　13. Order: Artiodactyla

14. Order: Hyracoidea
15. Order: Proboscidea
16. Order: Sirenia
17. Order: Cetacea
18. Order: Desmostylia
19. Order: Embrithopoda
20. Order: Notoungulata
21. Order: Pyrotheria
22. Order: Astrapotheria
23. Order: Tillodontia
24. Order: Litopterna
25. Order: Pantodonta
26. Order: Dinocerata
27. Order: Pholidota
28. Order: Tubulidentata
29. Order: Taeniodonta
30. Order: Xenarthra

Selected Bibliography

This listing contains only the most important volumes used during preparation of this work and a selection of textbooks and handbooks that provide further information.

Bellairs, A. d'A., J. Attridge: Reptiles. Hutchinson Univ. Libr., London 1975

Berg, L. S.: System der rezenten und fossilen Fischartigen und Fische. VEB Deutscher Verlag der Wiss., Berlin 1958

Brasier, M.D.: Microfossils. Allen & Unwin, London 1980

Clarkson, E. N. K.: Invertebrate Palaeontology and Evolution. Allen & Unwin, London 1979

Colbert, E. H.: Wandering Lands and Animals. Dutton, New York 1973

Colbert, E. H.: Evolution of the Vertebrates. Wiley, New York 1980

Fairbridge, R. W., D. Jablonski: The Encyclopedia of Paleontology. Dowden, Hutchinson & Ross, Stroudsburg, Pennsylvania 1979

Grassé, P. P.: Traité de Zoologie (a still uncompleted handbook, 17 volumes). Masson, Paris 1948ff

Gregory, W. K.: Evolution Emerging. Vol. I–II. Macmillan, New York 1951

Hadorn, E., R. Wehner: Allgemeine Zoologie, 20. Aufl. Thieme, Stuttgart 1978

Haq, B. U., A. Boersma: Introduction to Marine Micropaleontology. Elsevier, New York 1978

Hyman, L. H.: The Invertebrates. Vol. I–V. McGraw-Hill, New York 1940–1959

Jarvik, E.: Basic Structure and Evolution of Vertebrates. Vol. I–II. Academic Press, London 1980

Kuhn, O.: Handbuch der Paläoherpetologie (a still uncompleted, 19-part handbook). Fischer, Stuttgart 1969ff

Kuhn-Schnyder, E.: Geschichte der Wirbeltiere. Schwabe, Basel 1953

Lehmann, U.: Paläontologisches Wörterbuch. Enke, Stuttgart 1977

Lehmann, U., G. Hillmer: Wirbellose Tiere der Vorzeit. Enke, Stuttgart 1980

Lillegraven, J. A., Z. Kielan-Jaworowska, W. A. Clemens: Mesozoic Mammals. Univ. Calif. Press, Los Angeles 1979

Meglitsch, P. A.: Invertebrate Zoology. Oxford Univ. Press, New York 1967

Moore, R. C.: Treatise on Invertebrate Paleontology (a still uncompleted handbook, many volumes). Univ. Kansas Press, Lawrence, Kansas 1953ff

Moore, R. C., C. G. Lalicker, A. G. Fischer: Invertebrate Fossils. McGraw-Hill, New York 1952

Moy-Thomas, J. A., R. S. Miles: Palaeozoic Fishes. Chapman & Hall, London 1971

Müller, A. H.: Lehrbuch der Paläozoologie. Bd. I: Allgemeine Grundlagen, VEB Fischer, Jena 1976

Müller, A. H.: Lehrbuch der Paläozoologie. Bd. II: Invertebraten, VEB Fischer, Jena 1963–1978

Müller, A. H.: Lehrbuch der Paläozoologie. Bd. II: Invertebraten, VEB Fischer, Jena 1963–1978

Müller, A. H.: Lehrbuch der Paläozoologie. Bd. III: Vertebraten, VEB Fischer, Jena 1966–1970

Norman, J. R.: Die Fische. Parey, Hamburg 1966

Piveteau, J.: Traité de Paléontologie, Vol. I–VII. Masson, Paris 1952–1966

Piveteau, J., J.-P. Lehmann, C. Dechaseaux: Précis de Paléontologie des Vertébrés. Masson, Paris 1978

Raup, D. M., S. M. Stanley: Principles of Paleontology. Freeman, San Francisco 1978

Remane, A., V. Storch, U. Welsch: Systematische Zoologie: Stämme des Tierreichs. Fischer, Stuttgart 1976

Romer, A. S.: Review of the labyrinthodontia. Bull. Mus. Comp. Zool. Harvard 99/1 (1947)

Romer, A. S.: Osteology of the Reptiles. Univ. Chicago Press, Chicago 1956

Romer, A. S.: Vertebrate Paleontology. Univ. Chicago Press, Chicago 1966

Romer, A. S.: Notes and Comments on Vertebrate Paleontology. Univ. Chicago Press, Chicago 1968

Romer, A. S.: Vergleichende Anatomie der Wirbeltiere, Parey, Hamburg 1971

Schultze, H.-P.: Handbook of Paleoichthyologie (a still uncompleted, 10-part handbook). Fischer, Stuttgart 1978ff

Simpson, G. G.: The Major Features of Evolution. Columbia Univ. Press, New York 1953

Starck, D.: Vergleichende Anatomie der Wirbeltiere. Bd. I: Theoretische Grundlagen, Bd. II: Das Skeletsystem, Bd. III: Organe des aktiven Bewegungsapparates etc. Springer, Berlin 1978, 1979, 1982

Storer, T. I.: General Zoology. McGraw-Hill, New York 1951

Tasch, P.: Paleobiology of the Invertebrates. Wiley, New York 1973

Thenius, E.: Stammesgeschichte der Säugetiere (einschliesslich der Hominiden). In: Handb. Zool., 8/2. Gruyter, Berlin 1969

Thenius, E.: Die Evolution der Säugetiere. Fischer, Stuttgart 1979

Thenius, E.: Grundzüge der Faunen- und Verbreitungsgeschichte der Säugetiere. Fischer, Jena 1980

Weber, M.: Die Säugetiere, Bd. I: Anat. Teil, Bd. II: System Teil. Fischer, Jena 1927, 1928

Young, J. Z.: The Life of Vertebrates. Clarendon, London 1962

Ziegler, B.: Einführung in die Paläobiologie, Teil 1.: Allgemeine Paläontologie. Schweizerbart, Stuttgart 1972

Ziswiler, V.: Spezielle Zoologie: Wirbeltiere. Bd. I: Anamnia, Bd. II: Amniota. Thieme, Stuttgart 1976

v. Zittel, K. A.: Grundzüge der Paläontologie. II. Abteilung: Vertebrata. R. Oldenbourg, München 1923

v. Zittel, K. A.: Grundzüge der Paläontologie. I. Abteilung: Invertebrata. R. Oldenbourg, München 1924

Index of Genera and Species

Italics refer to figure legends.

Subject Index